HITE 6.0 培养体系

HITE 6.0全称厚溥信息技术工程师培养体系第6版，是武汉厚溥企业集团推出的“厚溥信息技术工程师培养体系”，其宗旨是培养适合企业需求的IT工程师，该体系被国家工业和信息化部人才交流中心鉴定为国家级计算机人才评定体系，凡通过HITE课程学习成绩合格的学生将获得国家工业和信息化部颁发的“全国计算机专业人才证书”，该体系教材由清华大学出版社全面出版。

HITE 6.0是厚溥最新的职业教育课程体系，该职业体系旨在培养移动互联网开发工程师、智能应用开发工程师、企业信息化应用工程师、网络营销技术工程师等。它的独特之处在于每年都要根据技术的发展进行课程的更新。在确定HITE课程体系之前，厚溥技术中心专业研究员在IT领域和一些非IT公司中进行了广泛的行业调查，以了解他们在目前和将来的工作中会用到的数据库系统、前端开发工具和软件包等应用程序，每个产品系列均以培养符合企业需求的软件工程师为目标而设计。在设计之前，研究员对IT行业的岗位序列做了充分的调研，包括研究从业人员技术方向、项目经验和职业素质等方面的需求，通过对面向学生的自身特点、行业需求与现状以及实施等方面的详细分析，结合厚溥对软件人才培养模式的认知，按照软件专业总体定位要求，进行软件专业产品课程体系设计。该体系集应用软件知识和多领域的实践项目于一体，着重培养学生的熟练度、规范性、集成和项目能力，从而达到预定的培养目标。整个体系基于ECDIO工程教育课程体系开发技术，可以全面提升学生的价值和学习体验。

一、移动互联网开发工程师

在移动终端市场竞争下，为赢得更多用户的青睐，许多移动互联网企业将目光瞄准在应用程序创新上。如何开发出用户喜欢，并能带来巨大利润的应用软件，成为企业思考的问题，然而这一切都需要移动互联网开发工程师来实现。移动互联网开发工程师成为求职市场的宠儿，不仅薪资待遇高，福利好，更有着广阔的发展前景，倍受企业重视。

移动互联网企业对Android和Java开发工程师需求如下：

已选条件:	Java(职位名)	Android(职位名)
共计职位:	共51014条职位	共18469条职位

1. 职业规划发展路线

Android				
★	★★	★★★	★★★★	★★★★★
初级Android 开发工程师	Android 开发工程师	高级Android 开发工程师	Android 开发经理	移动开发 技术总监
Java				
★	★★	★★★	★★★★	★★★★★
初级Java 开发工程师	Java 开发工程师	高级Java 开发工程师	Java 开发经理	技术总监

2. 素质能力提升路径

1 大学生	2 大学生活	3 学习习惯	4 职业目标	5 沟通表达	6 自我管理
12 准职业人	11 职业路线	10 求职技能	9 就业意识	8 融入团队	7 形象礼仪

3. 专业技能提升路径

1 大学生	2 计算机基础	3 编程基础	4 软件工程	5 数据库	6 网站技术
12 准职业人	11 产品规划	10 项目技能	9 高级应用	8 APP开发	7 基础应用

4. 项目介绍

(1) 酒店点餐助手

(2) 音乐播放器

二、智能应用开发工程师

随着物联网技术的高速发展，我们生活的整个社会智能化程度将越来越高。在不久的将来，物联网技术必将引起我国社会信息的重大变革，与社会相关的各类应用将显著提升整个社会的信息化和智能化水平，进一步增强服务社会的能力，从而不断提升我国的综合竞争力。 智能应用开发工程师未来将成为热门岗位。

智能应用企业每天对.NET开发工程师需求约15957个需求岗位(数据来自51job)：

已选条件:	.NET(职位名)
共计职位:	共15957条职位

1. 职业规划发展路线

★	★★	★★★	★★★★	★★★★★
初级.NET 开发工程师	.NET 开发工程师	高级.NET 开发工程师	.NET 开发经理	技术总监
★	★★	★★★	★★★★	★★★★★
初级 开发工程师	智能应用 开发工程师	高级 开发工程师	开发经理	技术总监

2. 素质能力提升路径

1 大学生	2 大学生活	3 学习习惯	4 职业目标	5 沟通表达	6 自我管理
12 准职业人	11 职业路线	10 求职技能	9 就业意识	8 融入团队	7 形象礼仪

3. 专业技能提升路径

1 大学生	2 计算机基础	3 编程基础	4 软件工程	5 数据库	6 网站技术
12 准职业人	11 产品规划	10 项目技能	9 高级应用	8 智能开发	7 基础应用

4. 项目介绍

(1) 酒店管理系统

(2) 学生在线学习系统

三、企业信息化应用工程师

当前，世界各国信息化快速发展，信息技术的应用促进了全球资源的优化配置和发展模式创新，互联网对政治、经济、社会和文化的影响更加深刻，围绕信息获取、利用和控制的国际竞争日趋激烈。企业信息化是经济信息化的重要组成部分。

IT企业每天对企业信息化应用工程师需求约11248个需求岗位（数据来自51job）：

已选条件：	ERP实施(职位名)
共计职位：	共11248条职位

1. 职业规划发展路线

初级实施工程师	实施工程师	高级实施工程师	实施总监
信息化专员	信息化主管	信息化经理	信息化总监

2. 素质能力提升路径

1 大学生	2 大学生活	3 学习习惯	4 职业目标	5 沟通表达	6 自我管理
12 准职业人	11 职业路线	10 求职技能	9 就业意识	8 融入团队	7 形象礼仪

3. 专业技能提升路径

1 大学生	2 计算机基础	3 编程基础	4 软件工程	5 数据库	6 网站技术
12 准职业人	11 产品规划	10 项目技能	9 高级应用	8 实施技能	7 基础应用

4. 项目介绍

(1) 金蝶K3

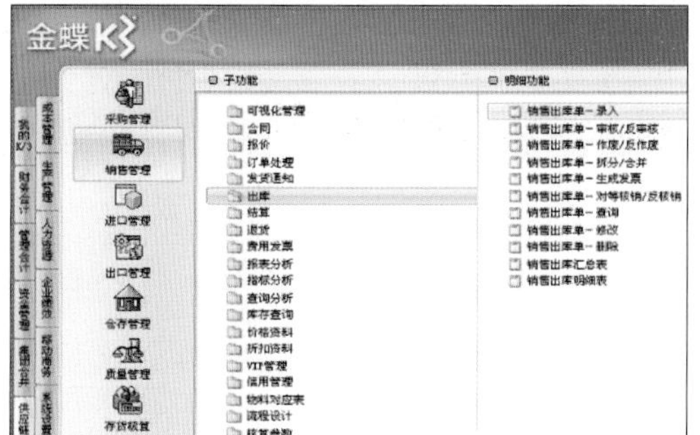

(2) 用友U8

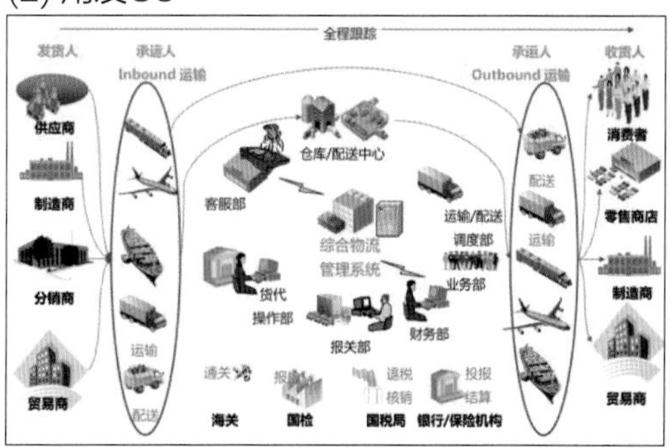

四、网络营销技术工程师

在信息网络时代，网络技术的发展和应用改变了信息的分配和接收方式，改变了人们生活、工作、学习、合作和交流的环境，企业也必须积极利用新技术变革企业经营理念、经营组织、经营方式和经营方法，搭上技术发展的快车，促进企业飞速发展。网络营销是适应网络技术发展与信息网络时代社会变革的新生事物，必将成为跨世纪的营销策略。

互联网企业每天对网络营销工程师需求约47956个需求岗位(数据来自51job)：

已选条件：	网络推广SEO(职位名)
共计职位：	共47956条职位

1. 职业规划发展路线

网络推广专员	网络推广主管	网络推广经理	网络推广总监
网络运营专员	网络运营主管	网络运营经理	网络运营总监

2. 素质能力提升路径

1 大学生	2 大学生活	3 学习习惯	4 职业目标	5 沟通表达	6 自我管理
12 准职业人	11 职业路线	10 求职技能	9 就业意识	8 融入团队	7 形象礼仪

3. 专业技能提升路径

1 大学生	2 计算机基础	3 编程基础	4 网站建设	5 数据库	6 网站技术
12 准职业人	11 产品规划	10 项目实战	9 电商运营	8 网络推广	7 网站SEO

4. 项目介绍

(1) 品牌手表营销网站

(2) 影院销售网站

HITE 6.0 软件开发与应用工程师

工信部国家级计算机人才评定体系

使用JSP开发企业级应用程序

武汉厚溥教育科技有限公司　编著

清華大學出版社
北　京

内 容 简 介

本书按照高等学校、高职高专计算机课程的基本要求，以案例驱动的形式来组织内容，突出计算机课程的实践性特点。本书共包括 11 个单元：Java Web 简介、JSP 简介、JSP 内置对象、Servlet 入门、会话管理和使用、会话和 Servlet 核心接口应用、表达式语言、JSP 标准标签库、过滤器、监听器和 MVC 模式。

本书内容安排合理，层次清楚，通俗易懂，实例丰富，突出理论和实践的结合，可作为各类高等院校、高职高专及培训机构的教材，也可供广大程序设计人员参考。

图书在版编目(CIP)数据

使用 JSP 开发企业级应用程序 / 武汉厚溥教育科技有限公司 编著. —北京：清华大学出版社，2019(2024.8重印)

(HITE 6.0 软件开发与应用工程师)

ISBN 978-7-302-52502-8

I. ①使… II. ①武… III. ①JAVA 语言—程序设计 IV. ①TP312.8

中国版本图书馆 CIP 数据核字(2019)第 043134 号

责任编辑：刘金喜
封面设计：贾银龙
版式设计：孔祥峰
责任校对：成凤进
责任印制：宋 林

出版发行：清华大学出版社
网 址：https://www.tup.com.cn, https://www.wqxuetang.com
地 址：北京清华大学学研大厦 A 座　　邮 编：100084
社 总 机：010-83470000　　邮 购：010-62786544
投稿与读者服务：010-62776969, c-service@tup.tsinghua.edu.cn
质 量 反 馈：010-62772015, zhiliang@tup.tsinghua.edu.cn

印 装 者：三河市君旺印务有限公司
经 销：全国新华书店
开 本：185mm×260mm　　**印 张**：20　　**插 页**：2　　**字 数**：474 千字
版 次：2019 年 4 月第 1 版　　**印 次**：2024 年 8 月第 5 次印刷
定 价：79.00 元

产品编号：082681-01

编委会

JSP是由Sun Microsystems公司倡导、许多公司参与一起建立的一种动态技术标准。在传统的网页HTML文件(*.htm、*.html)中加入Java程序片段(Scriptlet)和JSP标签，就构成了JSP网页。Java程序片段可以操纵数据库、重新定向网页以及发送E-mail等，实现建立动态网站所需要的功能。所有程序操作都在服务器端执行，网络上传送给客户端的仅是得到的结果，这样大大降低了对客户端浏览器的要求，即使客户端浏览器不支持Java，也可以访问JSP网页。

JSP全名为Java Server Pages，其根本是一个简化的Servlet设计，它实现了HTML语法中的Java扩展(以<%, %>形式)。JSP与Servlet一样，是在服务器端执行的。通常返回给客户端的就是一个HTML文本，因此客户端只要有浏览器就能浏览。Web服务器在遇到访问JSP网页的请求时，首先执行其中的程序段，然后将执行结果连同JSP文件中的HTML代码一起返回给客户端。插入的Java程序段可以操作数据库、重新定向网页等，以实现建立动态网页所需要的功能。

本书是“工信部国家级计算机人才评定体系”中的一本专业教材。“工信部国家级计算机人才评定体系”是由武汉厚溥教育科技有限公司开发，以培养符合企业需求的软件工程师为目标的IT职业教育体系。在开发该体系之前，我们对IT行业的岗位序列做了充分的调研，包括研究从业人员技术方向、项目经验和职业素质等方面的需求，通过对所面向学生的特点、行业需求的现状以及项目实施等方面的详细分析，结合我公司对软件人才培养模式的认知，按照软件专业总体定位要求，进行软件专业产品课程体系设计。该体系集应用软件知识和多领域的实践项目于一体，着重培养学生的熟练度、规范性、集成和项目能力，从而达到预定的培养目标。

本书共包括11个单元：Java Web简介、JSP简介、JSP内置对象、Servlet入门、会话管理和使用、会话和Servlet核心接口应用、表达式语言、JSP标准标签库、过滤器、监听器、MVC 模式。我们对本书的编写体系做了精心的设计，按照“理论学习—知识总结—上机操作—课后习题”这一思路进行编排。“理论学习”部分描述通过案例要达到的学习目标与涉及的相关知识点，使学习目标更加明确；“知识总结”部分概括案例所涉及的知识点，使知识点完整系统地呈现；“上机操作”部分对案例进行了详尽分析，通过完整的步骤帮助读者快速掌握该案例的操作方法；“课后习题”部分帮助读者理解章节的知识点。

本书在内容编写方面，力求细致全面；在文字叙述方面，注意言简意赅、重点突出；在案例选取方面，强调案例的针对性和实用性。

本书凝聚了编者多年来的教学经验和成果，可作为各类高等院校、高职高专及培训机构的教材，也可供广大程序设计人员参考。

本书由武汉厚溥教育科技有限公司编著，由翁高飞、张江城、许婧祺、丁文、郭彦、曾庆毅等多名企业实战项目经理编写。本书的编写团队长期从事项目开发和教学实施，并且对当前高校的教学情况非常熟悉，在编写过程中充分考虑到不同学生的特点和需求，加强了项目实战方面的教学。本书编写过程中，得到了武汉厚溥教育科技有限公司各级领导的大力支持，在此对他们表示衷心的感谢。

参与本书编写的人员还有：黔东南民族职业技术学院的杨琦、唐菲、李凌霄、宋开旭、田新宇、谢凌雁，武汉厚溥教育科技有限公司的李盛、夏小山等。限于编写时间和编者的水平，书中难免存在不足之处，希望广大读者批评指正。

服务邮箱：wkservice@vip.163.com。

编　者

2018 年 10 月

目录

单元一

Java Web 简介

课程目标

- 理解 Web 应用程序的概念
- 理解 Web 服务器的作用
- 掌握 Java Web 应用程序的目录结构

简 介

学习本书之前，我们都已经学习了面向对象的 Java 语法、JDBC 等相关技术。实际上，Java 语言更多地应用在 Java Web 开发中。本章通过介绍基于 Java 语言的三大开发平台、C/S 和 B/S 两种开发模式、三层体系结构等概念，让大家对 Java Web 技术有一定的了解，最后学习 Tomcat 服务器的使用。

1.1 Java Web 概述

随着计算机技术的快速发展，网络的出现改变了人们的生活方式，特别是近十年来，互联网更是取得了令人惊叹的发展速度。中国互联网的发展引领着整个世界的变革，利用网络进行的商务交易、在线办公、业务办理、数据的高速交互，都归功于 Web 应用所带来的便捷。

本书及后续的 Java 课程中，我们将深入学习 Java EE。Java EE 分为组件和服务两个部分，其中组件包括 Servlet、JSP、EJB；服务包括 JNDI、RMI-IIOP、JMS、JavaMAIL、Java Connector、JIDL、JAAS、JTA 等。

Java Web 其实就是前面提到的 Servlet 和 JSP 以及相关技术。Servlet 是服务器端的 Java 程序，JSP(Java Server Page)是服务器端的页面。如前所述，Servlet 和 JSP 是大部分开发人员必须掌握的，主要是针对基于 Java 语言的 Web 开发，所以我们将在接下来的学习中，专门学习这两项技术。

1.2 Java 开发软件架构

我们知道利用 Java SE 技术可以开发桌面应用程序，利用 HTML 网页技术和 JavaScript 技术可以开发网页，这些技术和我们将要接触的 Java Web 有关系吗？在了解这些内容之前，首先我们了解一下不同的开发模式。

1.2.1 C/S 和 B/S 的概念

在目前的企业级开发中，常用的开发模式有两种：C/S 和 B/S。

C/S，即 Client/Server。虽然从定义上，C/S 模式是指任何将事务处理分开进行的网络系统，但绝大多数的 C/S 应用系统都是 C/S数据库系统，充当服务器的是大型数据库系统，如 SQL Server、Oracle、MySql 等。

例如，使用 Java Swing 开发的数据库管理系统，Java 程序安装在各个不同的客户端，数据库则只存在于一台服务器主机上。多个用户在各自的主机上，使用该程序操作同一个数据库，如图 1-1 所示。

B/S，即 Browser/Server。B/S 模式是指软件的主体部分都在服务器端，用户只需使用浏览器发出请求，由服务器执行对请求动作的运算，并将最终运算结果发回到客户端即可。

例如，我们经常浏览的网站、论坛，客户机上只需要有浏览器，不需要安装任何客户端程序，所有的程序都在 Web 服务器上。用户通过浏览器向服务器发送请求，浏览器根据不同用户发出的不同请求，执行不同的程序，将计算出的结果显示到各个用户的浏览器上。在此说明一下，服务器包括数据库服务器和 Web 服务器，用户请求的是 Web 服务器，通过 Web 服务器上的程序再去对数据库进行操作，如图 1-2 所示。

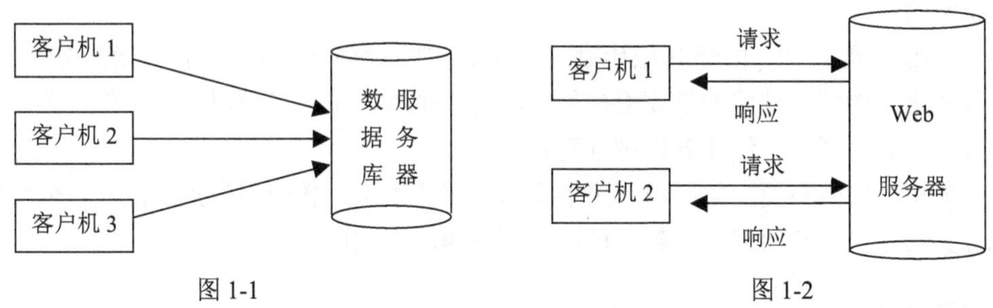

图 1-1　　图 1-2

1.2.2　C/S 和 B/S 的区别

我们将从以下几个方面分别阐述这两种开发模式的区别。

1. 数据安全性

由于 C/S 结构软件的数据分布特性，客户端所发生的火灾、盗抢、地震、病毒、黑客等都成了可怕的数据杀手。另外，对于集团级的异地软件应用，C/S 结构的软件必须在各地安装多个服务器，并在多个服务器之间进行数据同步。如此一来，每个数据点上的数据安全都影响了整个应用的数据安全。所以，对于集团级的大型应用来讲，C/S 结构软件的安全性是无法令人接受的。

对于 B/S 结构的软件来讲，由于其数据集中存放于总部的数据库服务器上，客户端不保存任何业务数据和数据库连接信息，也无须进行数据同步，所以这些安全问题自然也就不存在了。

2. 数据一致性

在 C/S 结构软件的解决方案里，异地经营的大型集团都采用各地安装区域级服务器，然后再进行数据同步的模式。这些服务器每天必须同步完毕之后，总部才可得到最终的数据。由于局部网络故障会造成个别数据库不能同步，即使同步上来，各服务器也不是一个时点上的数据，数据将永远无法一致，不能用于决策。

对于 B/S 结构的软件来讲，其数据是集中存放的，客户端发生的每一笔业务单据都直接进入到中央数据库，不存在数据一致性的问题。

3. 数据实时性

在集团级应用里，C/S 结构不可能随时随地看到当前业务的发生情况，我们看到的都是事后数据。B/S 结构则不同，它可以实时地显示当前所发生的所有业务，方便员工快速决策，有效地避免了企业的损失。

4. 可维护性

企业的业务流程、业务模式并不是一成不变的，随着企业的不断发展，必然会不断调整。软件供应商提供的软件也不是完美无缺的，所以，对已经部署的软件产品进行维护、升级是正常的。

对于 C/S 结构软件，由于其应用是分布的，需要对每一个客户端进行程序安装，所以，即使非常小的程序缺陷都需要很长的时间重新部署，重新部署时，为了保证各程序版本的一致性，必须暂停一切业务再进行更新。

B/S 结构的软件不同，其应用都集中于总部服务器上，各应用节点并没有任何程序，一个地方更新则全部应用程序更新，可以做到快速服务响应。

5. 响应速度

C/S 结构中，由于客户端实现与数据库服务器是直接相连的，所以响应速度非常快。

B/S 结构中，Web 应用程序动态刷新，所以响应速度明显变慢。

6. 界面设计

C/S 结构中，客户操作界面个性化，具有简单、直观、方便的特点，可以根据客户要求订制，充分满足客户个性化需求。

B/S 结构中，个性化特点明显降低，以鼠标为基本操作方式，无法实现快速操作的要求。

7. 服务器负载

C/S 结构中，客户端和服务器端都能处理任务，虽然对客户机要求较高，但是可以大大减少服务器的压力。

B/S 结构中，绝大部分工作由服务器承担，使得服务器负担很重。

以上是从几个方面对 C/S、B/S 结构进行比较，可以看出它们各有优缺点，不同的需求适合不同的开发模式。但是我们也能看出，随着网络越来越普及，PDA、智能手机、智能家电等计算机以外的上网方式发展迅速，也加速了 B/S 模式的发展。

1.3 Java Web 开发分层模型

在学习 Java 面向对象的过程中，我们学会了构建模型的思维方式，而在 Java Web 应用开发中，为了使程序具有良好的可扩展性、可维护性和可读性，我们将学习分层模型。一般可以分为如下几层。

1. 业务逻辑层

业务逻辑层由一系列的业务逻辑对象组成，封装一些方法通过调用数据存储层，向控制器层提供服务。这些业务逻辑方法可能仅仅用于暴露 Domain 对象所实现的业务逻辑方法，也可能是依赖 DAO 组件实现的业务逻辑方法。

2. 控制器层

控制器层由一系列的控制器组成，这些控制器用于拦截用户的请求，然后调用业务逻辑组件的业务逻辑方法，去处理用户的一系列请求，并根据处理结果转发到不同的表现层组件。

3. 前端层

前端层是由一系列的 JSP 页面、jQuery、AngularJs 等各种前端框架组成，负责收集用户的请求，并显示处理结果。

下面我们以图解的方式来展示 Web 分层架构，如图 1-3 所示。每一层之间都以松耦合的方式，方便对应用进行扩展，从上至下，上面组件的实现依赖下面组件的功能；从下至上，下面组件支持上面组件的实现。

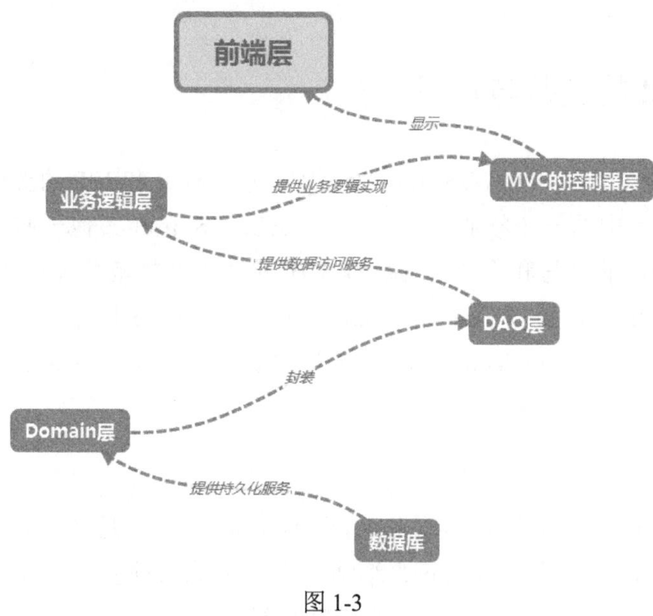

图 1-3

1.4 Web 服务器简介

前面我们反复提到了一个重要概念——Web 服务器，到底什么是 Web 服务器？它是如何工作的？既然是服务器，都有哪些种类？如何使用？

1.4.1 Web 服务器

Web 服务器，即在网络中为实现信息发布、资料查询、数据处理等诸多应用而搭建的基本平台的服务器。有时，我们也称 Web 服务器为 Web 容器。常用的服务器有 Tomcat、Resin、Weblogic、Websphere 等。在今后的学习中，我们将使用 Tomcat 服务器。

在 Web 应用中的处理可分为三个步骤：第一步，Web 浏览器向一个特定的 Web 服务器发出 Web 页面请求；第二步，Web 服务器接收到 Web 页面请求后，寻找所请求的 Web 页面，执行相应的功能；第三步，将用户请求的最后结果以 Web 页面形式发送到客户的 Web 浏览器，原理如图 1-4 所示。

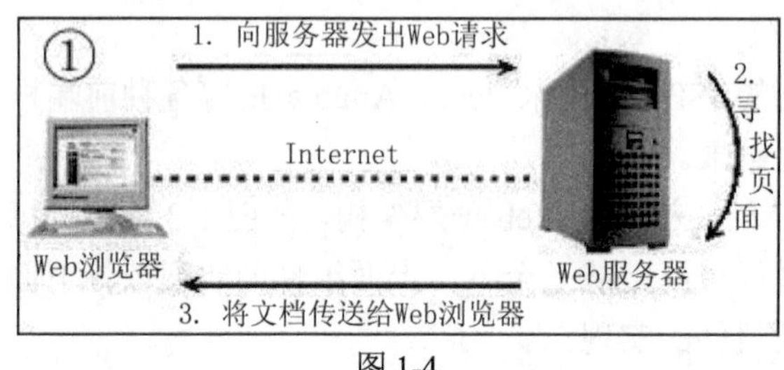

图 1-4

1.4.2 Tomcat 的安装与介绍

Tomcat 是一个免费的开源 Web 服务器，提供对 Servlet 和 JSP 的支持。它是 Apache 基金会 Jakarta 项目中的一个核心项目，由 Apache、Sun 和其他一些公司及个人共同开发。由于有了 Sun 的参与和支持，最新的 Servlet 和 JSP 规范总能在 Tomcat 中得到体现。本书介绍最新的稳定版本 Tomcat 8.5.32，支持 Servlet 3.1 规范及 JSP 2.3，下载网站是 http://tomcat.apache.org/。Tomcat 服务器非常适合学习 Java Web 开发技术的初学者使用。

1. 安装 Tomcat

在安装 Tomcat 之前，必须先安装 JDK，因为 Tomcat 本身是纯 Java 程序，需要 JVM 才能运行，此外 JSP 页面需要 javac 来编译运行，因此必须安装 JDK。本书中使用的 JDK 版本是 1.8.0_161。Apache 的官方站点提供基于 Windows 的安装版本，按步骤安装即可。

2. Tomcat 目录结构

成功安装 Tomcat 后，会产生如图 1-5 所示的目录和文件。

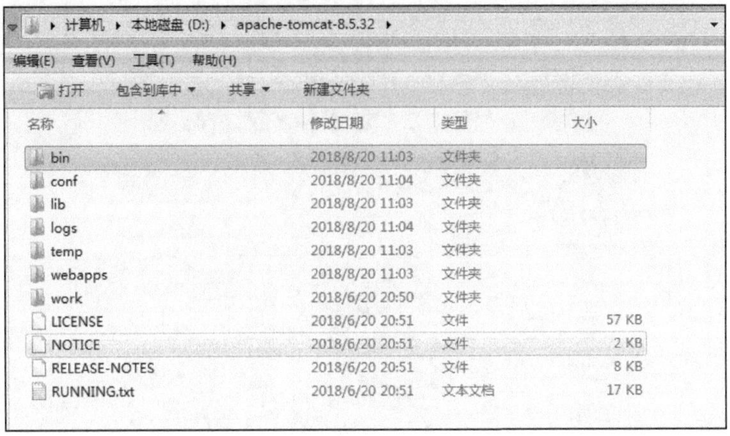

图 1-5

表 1-1 列出了每个目录的作用。

表 1-1

目　录	作　用
bin	存放启动 Tomcat 的可执行文件
conf	存放 Tomcat 服务器的各种配置文件，最重要的配置文件是 server.xml
lib	存放服务器需要的各种 jar 文件
logs	存放服务器的日志文件
temp	存放临时文件
webapps	存放发布 Web 应用的目录
work	Tomcat 把 JSP 生成的 Servlet 放在该目录下

这里我们需要先了解 webapps 目录，这个目录是用来部署所有 Web 应用程序的。

3. Tomcat 服务的启动和关闭

将编写好的 Web 应用程序放到 webapps 目录中，启动 Tomcat 服务器，用户就可以通过浏览器访问该程序了。启动 Tomcat 首先需要配置一个环境变量 JAVA_HOME，指向 JDK 的安装路径，环境变量配置好后，进入 Tomcat 安装路径下的 bin 目录，双击 startup.bat 文件即可启动 Tomcat 服务。服务启动后，打开浏览器，在地址栏中输入 http://localhost:8080，将显示如图 1-6 所示的界面，表明服务器成功启动。关闭服务启动命令窗口，即可关闭服务。此时若再次访问 http://localhost:8080，不会出现图 1-6 所示的界面。

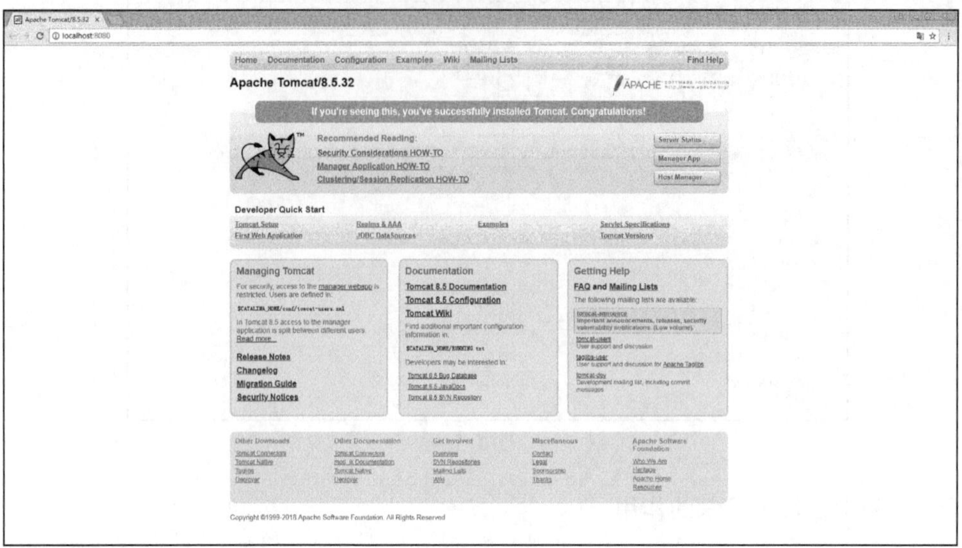

图 1-6

1.5 使用 Eclipse 开发第一个 Web 应用程序

1.5.1 Eclipse 集成 Tomcat

在项目开发中，为了简化操作，提高开发效率，会直接使用 Eclipse 工具集成 Tomcat 服务器，配置步骤如图 1-7 和图 1-8 所示。

本书中采用的 Eclipse 版本为 Oxygen.3 Release(4.7.3)，打开 Eclipse，单击菜单栏中的“Window→Preferences→Server→Runtime Environments”命令，单击“Add”按钮，如图 1-7 所示，添加 Apache Tomcat v8.5，然后选择其对应的本地路径，单击“Finish”按钮，如图 1-8 所示。

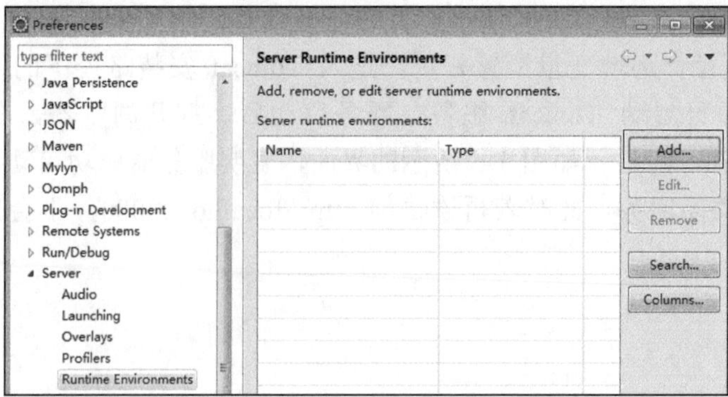

图 1-7

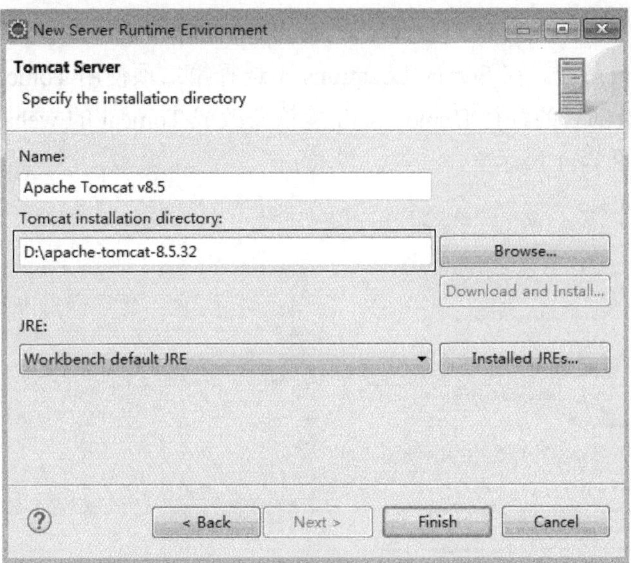

图 1-8

然后在 Eclipse 右下方界面，单击“Servers”选项，单击图 1-9 所示的链接。

图 1-9

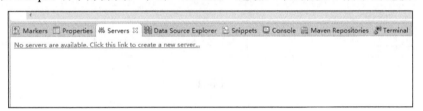

在弹出的窗口界面中可以看到刚刚配置好的 Tomcat，单击“Finish”按钮，效果图如图 1-10 所示。

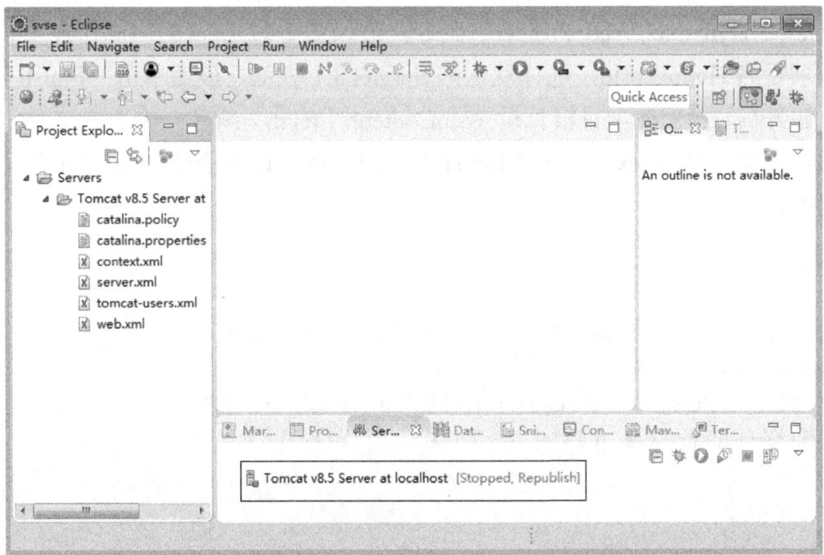

图 1-10

双击图 1-10 中的 Tomcat v8.5 Server at localhost 图标，进一步设置服务器路径以及项目在服务器下的发布路径。在 Server Locations 下选择第二项使用 Tomcat 的安装目录做完部署的位置，并修改部署的路径 Deploy path(建议修改成 Tomcat 的 webapps 目录)，最后按 Ctrl+S 键保存，如图 1-11 所示。

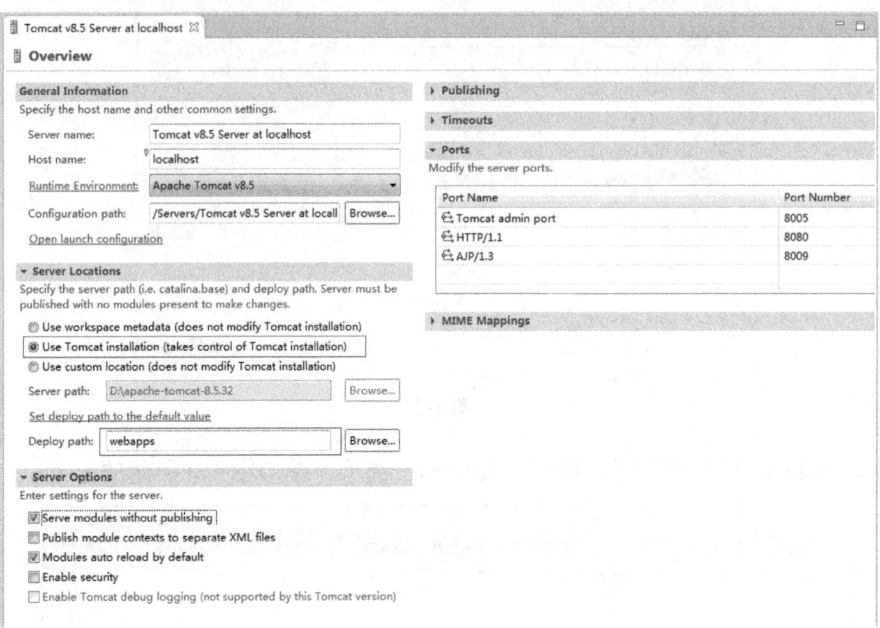

图 1-11

1.5.2 使用 Eclipse 开发 Web 应用

配置好 Tomcat 后，下面来开发第一个 Web 应用程序。启动 Eclipse，依次在菜单上单击“File→New→Dynamic Web Project”命令，进入图 1-12 所示的界面。在 Project name 文本框输入项目名 HelloWorld，选择默认的 Servlet 3.1 规范，最后单击“Finish”按钮。

项目新建好后，我们在项目根目录 WebContent 下右击，新建一个 JSP 文件，如图 1-13 所示。然后输入文件名 index，单击“Finish”按钮完成。最后修改 index.jsp 页面代码，修改后的代码如示例 1-1 所示。

示例 1-1：

```
<%@ page language="java" contentType="text/html; charset=UTF-8"
    pageEncoding="UTF-8"%>
<!DOCTYPE html PUBLIC "-//W3C//DTD HTML 4.01 Transitional//EN"
"http://www.w3.org/TR/html4/loose.dtd">
<html>
<head>
<title>第一个 Web 应用程序</title>
</head>
```

```
<body>
欢迎进入 Java Web 世界!
</body>
</html>
```

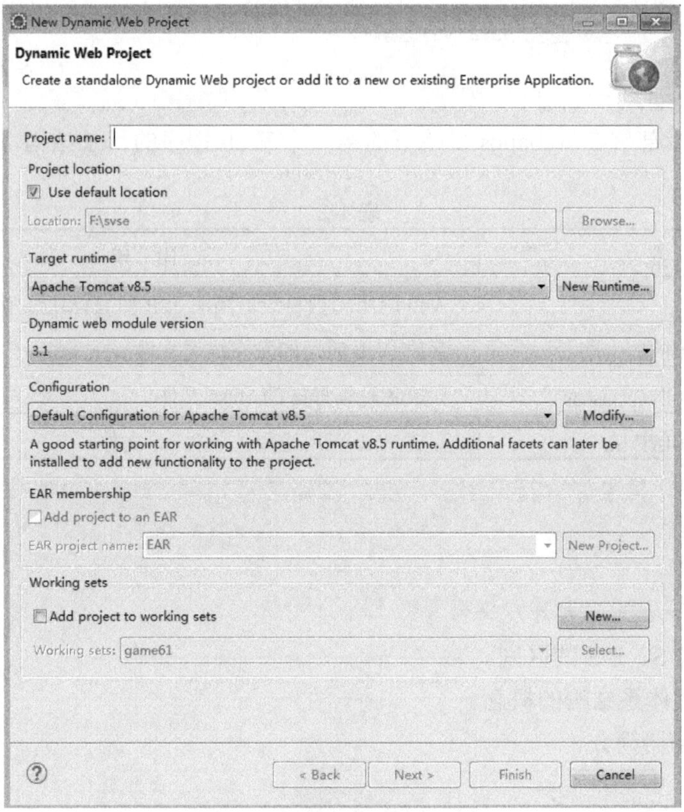

图 1-12

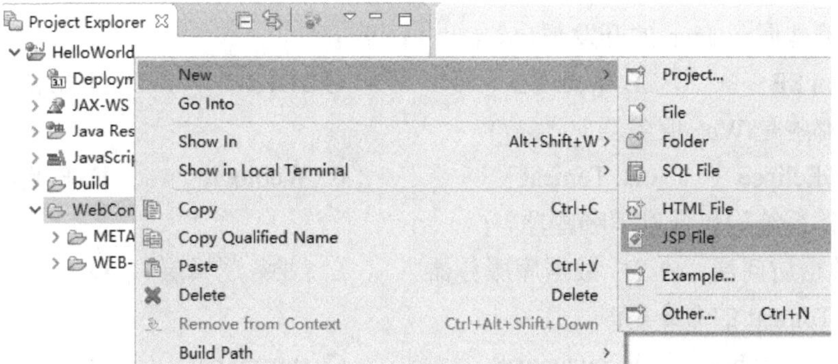

图 1-13

接下来，将项目部署到 Tomcat 中进行测试。右击项目名，选择“Run As→Run on Server”菜单项，在弹出的对话框中保持默认状态，单击“Finish”按钮，Web 项目部署完成。

若要停止 Web 服务器，可以单击 Servers 最右侧的■按钮，再次启动单击▶按钮即可。

我们可以打开浏览器，输入“localhost:8080/HelloWorld”，显示结果如图 1-14 所示。

图 1-14

让我们来看看部署在 webapps 目录下的第一个 Web 程序的目录结构，如表 1-2 所示。

表 1-2

目　录	作　用
/HelloWorld	Web 应用的根目录，所有 JSP 和 HTML 文件存放于此目录下
/HelloWorld/WEB-INF	存放 Web 应用的发布描述文件 web.xml
/HelloWorld/WEB-INF/classes	存放各种编译好的 class 文件
/HelloWorld/WEB-INF/lib	存放 Web 应用需要的各种 jar 文件

【单元小结】

- 介绍了 Java 的三大开发平台。
- 介绍了 C/S、B/S 的区别。
- 了解三层体系结构的概念。
- Web 服务器简介。

【单元自测】

1. 以下哪些属于 Java 的开发平台？(　　)

   A. Java SE　　B. Java EE　　C. JDK　　D. MyEclipse

2. 以下哪些是 Web 服务器？(　　)

   A. MyEclipse　　B. Tomcat　　C. Weblogic　　D. Eclipse

3. 三层体系结构包括以下哪些项？(　　)

   A. 数据访问层　　B. 数据库服务器　　C. 业务逻辑层　　D. 表示层

4. 启动 Tomcat 的命令是(　　)。

   A. shutdown.bat　　B. javac.exe　　C. startup.bat　　D. java.exe

5. 以下关于 B/S 和 C/S，哪些说法是不正确的？(　　)

   A. 在数据安全性方面，C/S 比 B/S 更有优势

   B. 在数据一致性方面，B/S 比 C/S 更有优势

   C. B/S 比 C/S 程序更加容易维护

单元二 JSP 简介

课程目标

- ▶ 了解 JSP 运行原理
- ▶ 掌握 JSP 脚本元素
- ▶ 掌握常用的 JSP 指令

简 介

Java Server Page(JSP)是基于Java的技术，用于创建可支持跨平台及跨Web服务器的动态网页。JSP可与微软的Active Server Page(ASP)相媲美，但JSP使用的是类似于HTML的标记和Java代码片段而不是VBScript。当使用不提供ASP本地支持的Web服务器(如Apache或Netscape服务器)时，就可以考虑使用JSP。虽然也可以为这些服务器加一个ASP附加软件模块，但是该类模块太昂贵了。现在Sun公司并不会收取你的JSP使用费(虽然将来可能要收)，况且用于Solaris、Linux以及Windows系统的组件都很容易获得。

2.1 JSP 运行原理

在单元一中，我们简单部署了一个HelloWorld项目，运行了index.jsp页面，可以通过浏览器访问url来查看页面内容，这是一个静态的页面，接下来我们在之前的基础上再编写一个jsp页面，如示例2-1所示。

示例 2-1：

```
<%@page import="java.util.Date"%>
<%@ page language="java" contentType="text/html; charset=UTF-8"
    pageEncoding="UTF-8"%>
<html>
<head>
<title>Insert title here</title>
</head>
<body>
欢迎来到厚溥学习 Java Web 知识，现在时间是：
<%out.println(new Date().toLocaleString()); %>
</body>
</html>
```

部署成功后，访问 http://localhost:8080/HelloWorld/demo1.jsp 页面，可以看到如图2-1所示的页面。

图 2-1

而此次demo1.jsp页面是一个含有动态数据的JSP页面。采用<%和%>将Java代码嵌入里面，可见JSP页面已不再需要Java类，似乎完全脱离了Java面向对象的特征。事实上，JSP的本质就是Servlet，每一个JSP页面就是一个Servlet实例。Servlet负责响应用户请求，在后面的章节中我们将系统学习 Servlet，在这里只需要将 Servlet 理解为具有一系列属性和方法的 Java 类即可。

下面我们通过图解来展示 JSP 的运行工作原理，如图 2-2 所示。

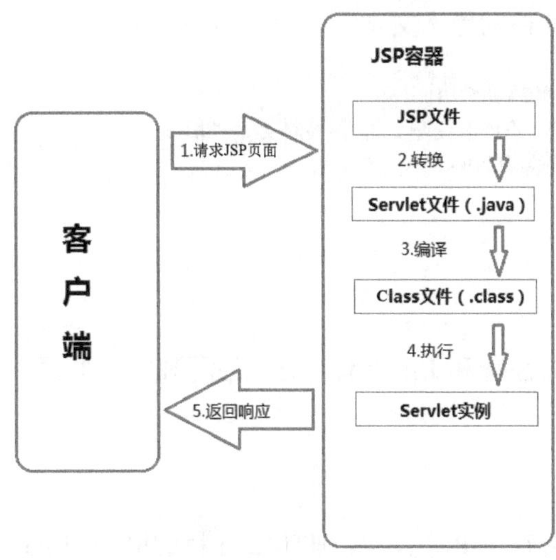

图 2-2

以上各步骤的含义如下。

(1) 客户端向服务器发送 JSP 页面请求。

(2) JSP 容器接收到请求后检索对应的 JSP 页面，如果该 JSP 页面是第一次被请求，则容器将此页面中的静态数据(HTML 文本)和动态数据(Java 脚本)全部转化成 Java 代码，使 JSP 文件翻译成一个 Java 文件，即 Servlet。

(3) 容器将翻译后的 Servlet 文件编译成字节码文件即.class。

(4) 编译后的字节码文件被加载到容器内存中执行，并根据用户的请求生成 HTML 格式的响应内容。

(5) 容器将响应内容返回到客户端。

注：如果访问的 JSP 页面已经被请求过，再次访问时，只要该 JSP 文件没有发生过改变，那么容器将直接调用装载好的字节码文件，不用再经历步骤(2)、(3)的过程，这样也能提高服务器的性能。

2.2 JSP 页面元素

JSP 的页面元素有静态内容、注释、脚本元素、指令、动作等。

2.2.1 JSP 静态内容

JSP 页面里的静态内容包括 HTML 的标签和文本，这些文本与 Java 代码和 JSP 无关，下面的代码片段就是 JSP 内的静态文本。

```
<head><title>第一个 JSP 页面</title></head>
<h1>今天开始学习 JSP，JSP 很简单，很容易掌握。</h1>
<h3>下面是 JSP 的一些特点</h3>
```

2.2.2 JSP 注释

JSP 注释分为 HTML 注释和 JSP 注释，这些注释独立于页面生成的内容的类型和页面所使用的脚本语言。

1. HTML 注释

HTML 注释可以包含在 JSP 文件中。HTML 注释的语法如下：

```
<!--  注释的内容 -->
```

这些注释将作为响应的一部分发回浏览器。由于它们是注释，所以不生成任何可见输出，但是最终用户可以通过浏览器的“查看”菜单的“源文件”选项看到这些注释。HTML 注释可以包含 JSP 表达式，而这些表达式生成的输出将作为注释的一部分出现在页面的响应中。

2. JSP 隐藏注释

JSP 隐藏注释只能通过原始的 JSP 文件来查看，当 JSP 作为响应返回到客户端时，客户端是看不到该注释的。JSP 隐藏注释的语法如下：

```
<%-- 注释的内容 --%>
```

JSP 容器会忽略注释的内容。在将页面转换为 Servlet 源代码时，将跳过出现在两个分隔符之间的所有内容。

3. Java 注释

Java 注释能通过原始的 JSP 文件和翻译后的 Servlet 源代码来查看，当 JSP 作为响应返回到客户端时，客户端是看不到该注释的。Java 注释的语法如下：

```
<%// 注释的内容  %>
```

被注释的内容会被翻译成 Servlet 源代码，但是不会编译为字节码文件，因此也不会被执行。

2.2.3 脚本元素

JSP 脚本元素使开发者能够直接将 Java 代码插入 JSP 页面中，这些 Java 代码将出现在由当前 JSP 页面生成的 Servlet 中。JSP 中提供了 3 种脚本元素：声明、表达式和代码段(Scriptlet)，如表 2-1 所示。

表 2-1

语 法	说 明	实 例
<%! 声明语句 %>	用于声明变量和方法	<%! String str="你好";%>
<%= 表达式 %>	将表达式的结果显示在页面中标签所在的位置	<%= obj.getName()%>
<% Java 代码 %>	Scriptlet 标签包含一个 Java 片段	<% int a = 10; for(int i=0;i<j;i++) out.print("数是" +i); %>

1. 声明

JSP 声明用来定义页面级变量，以保存信息或定义JSP页面的其余部分可能需要的方法。如果发现代码太多，最好把它们写成一个独立的 Java 类。声明的语法为：

```
<%! 声明语句 %>
```

在一个标记中会出现多个声明，但每一个声明都必须是完整的声明语句。

- 变量声明

JSP 页面里声明的变量将被转换和编译为对应 Servlet 类的属性。例如：

```
<%! int x = 5,y = 10; String str= "hello ";%>
```

这个声明能够在 Servlet 中创建 3 个属性，它们的属性名称分别为 x、y 和 str。也可以声明类属性，类属性则通过 static 关键字来定义，例如：

```
<%! public static int count = 0; %>
```

由于 Servlet 实例采取了单例模式，即在整个工程中只存在唯一实例，且常驻内存，所以被声明的变量是可以被多个请求之间共享的。如果任何一个请求修改此属性的值，则所有的请求都将拥有新值。

- 方法声明

JSP 页面里声明的方法将被转换和编译为对应的 Servlet 类的方法。例如，下面的代码片段中声明了判断奇偶数的方法。

```
<%!
    public String isEven(int num){
    if(num%2==0)
        return  "偶数";
    else
```

```
            return   "奇数";
    }
%>
```

多个方法定义可以出现在单个声明标记内，也可以同时声明变量和方法，如下所示：

```
<%!
//声明属性
private String username;
//声明 getter 方法
public String getName(){
    return username;
}
//声明 setter 方法
public void setName(String name ){
    this.username = name ; //使用 this 关键字来引用属性
}
%>
```

此声明中的 username 为一个类属性。这里需要注意的是，既然 username 是类的属性，我们就可以使用 this 关键字来引用这个属性。

2. 表达式

声明不能直接参与页面输出，页面输出的是动态内容生成的目标。有了 JSP 表达式，表达式结果会被转换成一个字符串，并且被直接包括在输出页面之内。表达式的语法如下：

```
<%= 表达式 %>
```

JSP 表达式可用于输出单个变量或某个计算的结果。例如，下面的表达式直接将圆周率 PI 的值输出到页面中：

```
<%= Math.PI %>
```

假设圆的半径变量 radius 已经存在，下面的表达式则用于输出以 radius 为半径的圆的面积：

```
<%= Math.PI * Math.pow(radius,2) %>
```

再例如，以上判断奇数和偶数方法的声明，可以使用下列表达式得到结果：

```
<%= isEven(320) %>
```

表达式可以返回简单类型变量的值，也可以返回类似 String 等 Java 对象。表达式的结果在被添加到页面的输出中之前被转换成字符串。要注意的是表达式后不能有“;”，否则就变成代码段 Scriptlet 了。

3. 代码段(Scriptlet)

在 JSP 页面中，可以在“<%”和“%>”标记之间直接嵌入任何有效的 Java 语言代码。

这样嵌入的代码片段称为 Scriptlet。如果要完成的任务比插入简单的表达式更加复杂，那么就要使用 Scriptlet。Scriptlet 的语法如下：

```
<% Java 代码 %>
```

代码段应该是 JSP 页面脚本语言的一个或多个有效的完整语句，代码段不经过任何转换，就被插入到 Servlet 的源代码中。下面的 JSP 中定义了 3 个程序片段：

```
<%
    String gender = "female";
    if(gender.equals("female")){
%>
    She is a girl.
<%
   }else{
%>
   He is a boy.
<%
 }
%>
```

上面的 JSP 代码输出结果为“She is a girl.”。在这里，if 语句由 3 段<% 和 %>代码构成，分段的 if 语句可以控制网页的输出结果。注意，这里“She is a girl”和“He is a boy”是 JSP 页面里的静态内容。

示例 2-2 所示是一个包含“<%!　%>”、“<%=　%>”和“<%　%>”标签的 JSP 完整例子 hitCounter.jsp。

示例 2-2：

```
<%@ page contentType="text/html;charset=UTF-8"%>
<html>
<head><title>第一个 JSP 页面</title></head>
<body>
<h1>
您是今天的第
<%!int hitcount=0;%> //注意这里也可以不给出初始化的值，为什么？
<% hitcount++;%>
<%=hitcount%>
个访问者！
</h1>
</body>
</html>
```

hitCounter.jsp 的输出结果如图 2-3 所示。

图 2-3

hitCounter.jsp 第一次被请求时输出的值为 1，如果用户不停地单击“刷新”按钮，counter 的值会逐渐递增。如果不同的用户从多个浏览器窗口来访问 hitCounter.jsp，counter 的值则不会从 1 开始，而是接着显示最后一次被访问时的值。想想，这是为什么？在单元四中我们会解答此问题。

2.2.4 指令

指令用于将有关页面的特殊处理信息传送到 JSP 容器。常用的 JSP 指令分为三类：第一类是 include 指令，用来在 JSP 文件转换成 Servlet 时引入其他文件；第二类是 page 指令，用来导入指定的类、自定义 Servlet 的超类等；第三类是 taglib 指令，其目的是让 JSP 开发者能够自己定义标签。

指令的语法如下：

```
<%@ 指令类型 指令属性%>
```

1. include 指令

include 指令用于 JSP 页面转换成 Servlet 时引入其他文件。该指令语法如下：

```
<%@ include file="URL" %>
```

这里所指定的 URL 是和发出引用指令的 JSP 页面相对的 URL，然而，与通常意义上的相对 URL 一样，可以利用以“/”开始的 URL 告诉系统把 URL 视为从 Web 服务器根目录开始。包含文件的内容也是 JSP 代码，即包含文件可以包含静态 HTML、脚本元素、JSP 指令和动作。

例如，许多网站在每个页面都有一个导航条。由于 HTML 框架存在不少问题，导航条往往用页面顶端或左边的一个表格制作，同一份 HTML 代码重复出现在整个网站的每个页面上。include 指令是实现该功能的非常理想的方法。使用 include 指令，开发者不必再把导航 HTML 代码复制到每个文件中，从而可以更轻松地完成维护工作。

由于 include 指令是在翻译阶段引入所包含的文件，因此如果导航条改变了，所有使用该导航条的 JSP 页面都必须重新翻译。如果导航条改动不频繁，并且希望包含操作具有尽可能高的效率，使用 include 指令是最好的选择。然而，如果导航条改动非常频繁，则可以使用<jsp:include>动作。<jsp:include>动作在出现对 JSP 页面请求的时候才会引用指定的文件。

2. page 指令

page 指令的作用是定义表 2-2 中的一个或多个属性，这些属性对大小写敏感。

表 2-2

属　性	值	默认值
import	类名或包名	无
language	脚本语言名称	"java"
contentType	MIME 类型和字符集	"text/html"
extends	类名	无
info	文本字符串	无
session	布尔值	"true"
buffer	缓冲器大小，或 none	"8kb"
autoFlush	布尔值	"true"
isThreadSafe	布尔值	"true"
errorPage	本地 URL	无
isErrorPage	布尔值	"false"

- import 属性

import 属性很常用，可以在 JSP 页面中使用类，而不必显式指定类包名称。要导入特定类，只需将其名称指定为 import 属性值即可，例如：

```
<%@ page import="java.util.List" %>
```

在使用同一个包中的多个类的情况下，也可以将整个包导入 JSP 页面，例如：

```
<%@ page import="java.util.*" %>
```

import 属性支持通过单一属性导入多个类或包，之间用逗号分隔即可，最后一个导入的类不需要使用“,”，否则会产生错误，例如：

```
<%@ page import="java.util.List, java.util.ArrayList, java.sql.* " %>
```

默认情况每个 JSP 页面都会自动导入以下 4 个包中的所有类：java.lang、javax.servlet、javax.servlet.http、javax.servlet.jsp。

- language 属性

language 属性指定使用的脚本语言。语法为：

```
<%@ page language="java" %>
```

- contentType 属性

contentType 属性指定输出的 MIME 类型，默认是 text/html。例如，下面的指令：

```
<%@ page contentType="text/html" %>
```

与下面的 Scriptlet 效果相同：

```
<% response.setContentType("text/html"); %>
```

为了让汉字能在页面上正常显示，通常按如下写：

```
<%@ page contentType="text/html;charset=GBK" %>
```

- extends 属性

```
<%@ page extends="package.class" %>
```

extends 属性指出将要生成的 Servlet 使用哪个超类。使用该属性应当十分小心，因为服务器可能已经在用自定义的超类。

- info 属性

```
<%@ page info="版权归 SVSE 所有" %>
```

info 属性定义一个可以通过 getServletInfo 方法提取的字符串。

- session 属性

```
<%@ page session="false " %>
```

默认值为 true，表明预定义变量 session 应该绑定到已有的会话，如果不存在已有的会话，则新建一个会话并绑定到 session 变量。如果取值 false，表明如果没有与请求相关联的会话，也不创建会话绑定到 session 变量。但是，此时如果访问变量 session，将导致 JSP 转换成 Servlet 时出错。

- buffer 属性

```
<%@ page buffer="12KB" %>
```

buffer 属性指定 JspWrite out 的缓存大小、默认值和服务器有关，但至少应该是 8 KB。如果要关闭缓冲输出，以便所有 JSP 内容立即传送给 HTTP 响应，则应该将此属性设置为 none，如下所示：

```
<%@ page buffer="none" %>
```

- autoFlush 属性

```
<%@ page autoFlush="true " %>
```

默认值为 true，表明如果缓存已满则刷新它。autoFlush 很少取 false 值，false 值表示如果缓存已满则抛出异常。如果 buffer="none"，autoFlush 不能取 false 值。

- isThreadSafe 属性

```
<%@ page isThreadSafe="true " %>
```

默认值为 true，代表 JSP 容器会以多线程方式运行 JSP 页面；如果设置为 false，JSP

容器会以单线程方式运行 JSP 页面。建议保持默认状态。

- errorPage 属性

```
<%@ page errorPage="error.jsp " %>
```

errorPage 属性指定一个 JSP 页面，所有未被当前页面捕获的异常均由该页面处理。

- isErrorPage 属性

```
<%@ page isErrorPage="true " %>
```

isErrorPage 默认值为 false，当值设置为 true 时，用于指定此 JSP 页面为处理异常错误的页面，并且此时通常无须再指定 errorPage 属性。

3. taglib 指令

taglib 指令用于通知 JSP 容器某个页面依赖于自定义标记库。标记库是可用于扩展 JSP 功能的自定义标记的集合。语法如下：

```
<%@ taglib uri="tag Library URI" prefix="tag Prefix" %>
```

其中，uri 属性的值表示库的标记描述符(TLD)文件的位置，而 prefix 属性指定将其作为前缀附加到页面的每一处库标记上的 XML 名字空间标识符。taglib 指令的详细用法，我们将在标准标签库的章节里详细介绍。

2.2.5 JSP 动作

JSP 中定义了若干个标准动作，它允许用户在页面中使用 JavaBean 组件和有条件地把页面的控制权转移到其他页面。下面的代码片段显示了在页面中使用 Userinfo 对象并把页面转到 show.jsp。

```
<jsp:useBean id="info" class="bean.Userinfo" scope="session"/>
<jsp:forward page="show.jsp" flush="true" />
```

【单元小结】

- JSP 是基于Java的技术，用于创建可支持跨平台及跨 Web 服务器的动态网页。
- JSP 提供了 3 种脚本元素：声明、表达式和代码段(Scriptlet)。
- 指令用于将有关页面的特殊处理信息传送到 JSP 容器。
- page 指令是最复杂的指令。

【单元自测】

1. JSP 默认的脚本语言为(　　)。

A. html　　B. jsp　　C. aspx　　D. java

2. page 指令中的 import 属性可以在页面中出现(　　)次。

A. 1　　B. 2　　C. 多　　D. 以上都不对

3. page 指令中的 contentType 属性的默认值是(　　)。

A. text/xml　　B. text/html　　C. text/plain　　D. image/gif

4. 下列关于 import 属性，说法错误的是(　　)。

A. import 属性能导入某一个特定的类

B. import 属性能导入某一个包中的所有类

C. import 属性能通过单一属性导入多个包或类

D. import 属性不能通过单一属性导入多个包或类

5. JSP 中的 3 种脚本元素是(　　)。

A. 声明　　B. Scriptlet　　C. 表达式　　D. 注释

【上机实战】

上机目标

- 掌握 JSP 脚本元素的使用。
- 掌握 JSP 指令的使用。

上机练习

◆ 第一阶段 ◆

练习 1：掌握 include 指令的使用以及脚本元素中的声明和代码段

【问题描述】

编写一个 JSP 程序，使用 include 指令包含另一个页面。

【问题分析】

编写两个 JSP 文件，在 main.jsp 中使用 include 指令将 info.jsp 包括到其中。

【参考步骤】

(1) 创建 Web 应用，取名 JSPWeb1。

(2) 创建 main.jsp 文件，代码如示例 2-3 所示。

示例 2-3：

```
<%@ page contentType="text/html; charset=UTF-8"%>
```

```
<html>
<body>
<h2>以下是 info.jsp 中的内容</h2>
<%@ include file="info.jsp"%>
</body>
</html>
```

(3) 创建 info.jsp 文件，代码如示例 2-4 所示。

示例 2-4：

```
<%@ page contentType="text/html; charset=UTF-8"%>
<html>
<body>
<%! String name="美国硅谷软件工程教育";%>
<marquee>
上大学,好工作,我选<%=name%>!
</marquee>
</body>
</html>
```

(4) 结果如图 2-4 所示。

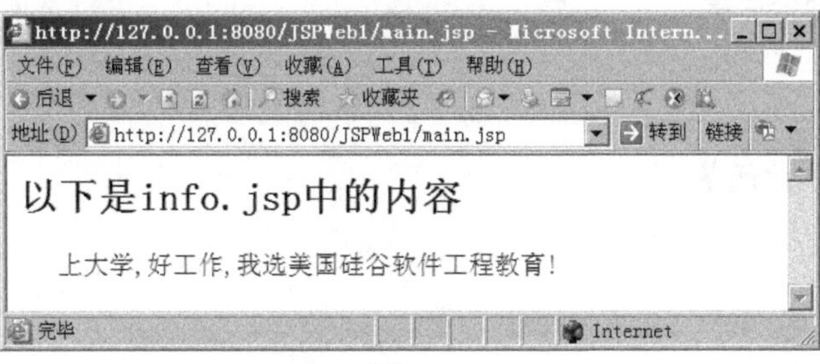

图 2-4

练习 2：掌握 page 指令的使用以及脚本元素中的声明和代码段

【问题描述】

在 JSP 页面上以“yyyy 年 MM 月 dd 日”的格式显示当前日期。

【问题分析】

(1) 使用 page 的 import 属性导入 java.util.Calendar 类。

(2) 在页面上分别输出年、月、日。

【参考步骤】

(1) 创建 date.jsp，代码如下。

```
<%@ page contentType="text/html;charset=UTF-8" import="java.util.Calendar"%>
<html>
<body>
```

```
<% Calendar now = Calendar.getInstance();%>
<h3>当前的日期为:
<%=now.get(Calendar.YEAR)%>年
<%=now.get(Calendar.MONTH)+1%>月
<%=now.getInstance().get(Calendar.DATE)%>日
</h3>
</body>
</html>
```

(2) 运行结果如图 2-5 所示。

当前的日期为: 2018年 8月 21日

图 2-5

◆ 第二阶段 ◆

练习 3：掌握脚本元素

【问题描述】

统计当前的用户访问量。

【问题分析】

(1) 在脚本元素声明中，定义 count 变量。

(2) 在脚本元素代码段中，递增 count。

(3) 使用脚本元素表达式显示 count 的值。

【拓展作业】

1. 在 JSP 页面中输出 10 行“我要学好 JSP！”。
2. 在 JSP 页面中显示当前日期是这一年的第几天。
3. 将个人爱好页面显示在个人信息的页面中。

单元三 JSP内置对象

课程目标

- ▶ 掌握输入/输出对象的使用
- ▶ 掌握作用域通信和控制对象的使用
- ▶ 掌握 Servlet 相关对象的使用
- ▶ 掌握错误处理

简 介

隐式对象是运行 JSP 的容器自动创建的 Java 类实例，允许与底层的 Servlet 环境交互。JSP 的一系列隐式对象和方法用于执行一些功能，如从客户端页面访问数据，发回数据，控制传输数据的缓冲，创建和访问 Cookie，在 page、session 和 application 等不同作用域级别进行通信和共享信息，管理作用域以及获得客户端/服务器环境信息等。

隐式对象可分为四大类：输入/输出对象、作用域通信和控制对象、Servlet 相关对象及错误处理对象，它们都可用在 JSP 页面的表达式和代码段中。

3.1 输入/输出对象

输入/输出隐式对象包括 request、response 和 out，常用于 JSP 页面的输入和输出。request 对象封装了客户端请求的信息，response 对象封装了服务器响应的结果，out 对象表示与响应关联的实际输出流。

3.1.1 request 对象

request 对象实现了 javax.servlet.http.HttpServletRequest 接口，在 request 对象中封装了客户请求的数据信息，如客户的请求方式、参数名和参数值、客户端正在使用的协议以及发出客户请求的远程主机信息等。request 对象还提供了直接以二进制方式读取客户请求数据流的 ServletInputStream 对象。

request 对象所提供的方法可以将它分为四大类：存储和获取属性方法，如表 3-1 所示；获取请求参数的方法，如表 3-2 所示；其他方法，如表 3-3 所示。这些方法都是在 javax.servlet.ServletRequest 接口中定义的方法。HttpServletRequest 接口中定义了获取 HTTP 请求标题头的方法，如表 3-4 所示。

表 3-1

方 法	说 明
void setAttribute(String name, Object value)	把 name 属性储存到 request 对象
Object getAttribute(String name)	返回 name 属性的值
void removeAttribute(String name)	从 request 对象中移除 name 属性

表 3-2

方 法	说 明
String getParameter(String name)	返回指定 name 的参数值
Enumeration getParameterNames()	以枚举类型返回所有的参数名称

(续表)

方　法	说　明
String [] getParameterValues(String name)	以字符数组类型返回所有指定 name 的参数值
Map getParameterMap()	以 java.util.Map 类型返回请求参数

表 3-3

方　法	说　明
String getProtocol()	返回使用的协议版本(例如 HTTP/1.1)
String getRemoteAddr()	返回用户的 IP 地址
String getRemoteHost()	返回用户的主机名称
int getRemotePort()	返回用户的主机端口
void setCharacterEncoding(String code)	设置请求正文的编码方式

表 3-4

方　法	说　明
String getContextPath()	返回请求 context 路径(即站点名称)
String getMethod()	返回 http 的方法(get、post、put 等)
String getQueryString()	返回请求 URI 后面包含的查询字符串
String getRequestedSessionId()	返回用户端的 session ID
String getRequestURI()	返回请求 URI，不包括请求的查询字符串
String getRequestURL()	返回请求的全部 URL，包括协议、服务器名字和端口、请求的 URL，但不包括请求的查询字符串
String getRemoteUser()	返回登录用户的名称
String getHeader(String name)	返回指定 name 的标题头
Enumeration getHeaderNames()	以枚举类型返回所有的标题头名称
Enumeration getHeaders(String name)	以枚举类型返回所有指定 name 的标题头
int getIntHeader(String name)	以整数类型返回指定 name 的标题头
long getDateHeader(String name)	以日期类型返回指定 name 的标题头
Cookie [] getCookies()	以 Cookie 数组类型返回与请求相关的所有 Cookie

示例 3-1 和示例 3-2 演示了上面方法的用法。sina.jsp 页面是 sina 邮箱的注册页面，register.jsp 是注册信息的确认页面，同时可以获得用户请求的相关信息。

示例 3-1：

```
<%@ page contentType="text/html; charset=UTF-8"%>
<html>
    <head><title>新浪邮箱注册</title></head>
    <body>
        <form action="register.jsp" method="post">
```

```
            创建你的新浪邮箱:
            <table>
                <tr>
                    <td align="right">选择你的新浪邮箱:</td>
                    <td>
                        <input type="text" name="uname">@
                        <select name="sel">
                        <option value="sina.cn">sina.com</option>
                        <option value="sina.com.cn">sina.com.cn</option>
                        </select>
                    </td>
                </tr>
                <tr>
                    <td align="right">密码:</td>
                    <td><input type="password" name="pwd"></td>
                </tr>
                <tr>
                    <td align="right">再次输入密码:</td>
                    <td><input type="password" name="rpwd"></td>
                </tr>
                <tr>
                    <td align="right">姓名:</td>
                    <td><input type="text" name="rname"></td>
                </tr>
                <tr>
                    <td align="right">性别:</td>
                    <td>
                    <input type="radio" name="sex" value="男">男 
                    <input type="radio" name="sex" value="女">女 
                    </td>
                </tr>
            </table>
            中国新浪直邮:<br>
            我愿意接受来自中国新浪的直邮，请选择(可多选)<br>
            <input type="checkbox" name="type" value="旅游">旅游 
            <input type="checkbox" name="type" value="休闲">休闲 
            <input type="checkbox" name="type" value="财经">财经 
            <input type="checkbox" name="type" value="汽车">汽车 
            <br>
            <input type="checkbox" name="type" value="音乐">音乐 
            <input type="checkbox" name="type" value="房产">房产 
            <input type="checkbox" name="type" value="家居">家居 
            <input type="checkbox" name="type" value="生活">生活 
            <br><br><input type="submit" value="注册新浪邮箱">
        </form>
    </body>
</html>
```

示例 3-2：

```
<%@ page contentType="text/html; charset=UTF-8"%>
<html>
  <head> <title>新浪邮箱注册信息</title></head>
  <body>
  <% request.setCharacterEncoding("UTF-8");%>
    请确认你的注册信息：<br>
    邮箱名:
    <%=request.getParameter("uname")%>@<%=request.getParameter("sel")%>
    <br>密 码：<%=request.getParameter("pwd") %><br>
    确定密码：<%=request.getParameter("rpwd") %><br>
    性 别:<%=request.getParameter("sex") %><br>
    选择的新浪直邮方式:<%
      String []types=request.getParameterValues("type");
      for(int i = 0 ; i <types.length ; i ++){
      %>
      <%=types[i]+"\t"%>
      <%} %>
      <br>
    <hr>
    来自客户端的信息：<br>
    用户请求的 URL:<%= request.getRequestURL()%><br>
    主机的网络地址:<%= request.getRemoteAddr()%><br>
    HTTP 请求方法:<%= request.getMethod()%><br>
    Accept 头包含 MIME 类型:<%= request.getHeader("Accept")%><br>
  </body>
</html>
```

运行 sina.jsp，结果如图 3-1 所示。

图 3-1

单击“注册新浪邮箱”按钮，结果如图 3-2 所示。

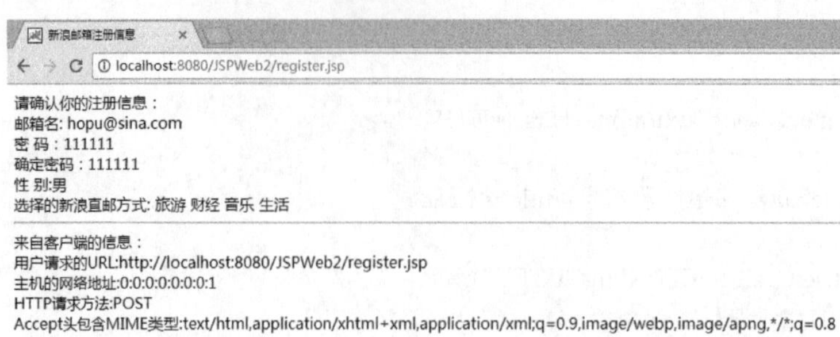

新浪邮箱注册信息
localhost:8080/JSPWeb2/register.jsp
请确认你的注册信息：
邮箱名: hopu@sina.com
密 码：111111
确定密码：111111
性 别:男
选择的新浪直邮方式: 旅游 财经 音乐 生活
来自客户端的信息：
用户请求的URL:http://localhost:8080/JSPWeb2/register.jsp
主机的网络地址:0:0:0:0:0:0:0:1
HTTP请求方法:POST
Accept头包含MIME类型:text/html,application/xhtml+xml,application/xml;q=0.9,image/webp,image/apng,*/*;q=0.8

图 3-2

3.1.2 response 对象

response 对象主要将 JSP 处理数据后的结果传回到客户端。response 对象实现了 javax.servlet.http.HttpServletResponse 接口，所提供的方法可以将它分为三大类：设定标题头的方法，如表 3-5 所示；设定响应状态码的方法，如表 3-6 所示；用来重写 URL 的方法，如表 3-7 所示。

表 3-5

方 法	说 明
void addCookie(Cookie cookie)	添加指定的 cookie 到响应对象
void addDateHeader(String name,long date)	添加指定的 name 和 long 类型值到标题头
void addHeader(String name,String value)	添加指定的 name 和 String 类型值到标题头
void addIntHeader(String name,int value)	添加指定的 name 和 int 类型值到标题头
void setDateHeader(String name,long date)	使用指定的 name 和 long 类型值设置标题头
void setHeader(String name,String value)	使用指定的 name 和 String 类型值设置标题头
void setIntHeader(String name,int value)	使用指定的 name 和 int 类型值设置标题头

表 3-6

方 法	说 明
void sendError(int sc)	向客户端发送定义的状态码
void sendError(int sc, string msg)	向客户端发送定义的状态码和错误信息
void setStatus(int sc)	设置响应的状态码

表 3-7

方 法	说 明
String encodeURL(String url)	重写指定的 URL，包含 session ID
String encodeRedirectURL(String url)	对使用 sendRedirect( )方法的 URL 重写

示例 3-3～示例 3-6 为使用 request 对象和 response 对象的方法实现登录页面的转向。

index.jsp 页面让用户输入用户名和密码，check.jsp 判断用户名和密码是否匹配，如果匹配，页面转到 welcome.jsp，否则，页面转到 error.jsp。

示例 3-3：

```
<%@ page contentType="text/html;charset=UTF-8"%>
<html>
<head>
<title>登录 index.jsp</title>
</head>
<body>
<form name="form1" method="post" action="check.jsp">
  <p>用户名:
    <input name="userName" type="text" >
  <p>密   码:
    <input name="password" type="password">
  <p>
<input type="submit" value="提交">
</form>
</body>
</html>
```

示例 3-4：

```
<%@ page contentType="text/html; charset=UTF-8"%>
<html>
<head><title>判断 check.jsp</title></head>
<body>
<%
String name = request.getParameter("userName");
String pwd = request.getParameter("password");
//把信息存放到 request 对象中
request.setAttribute("msg","张三丰");
   if(pwd.equals("123")&&name.equals("bush")){
      //页面转向方式 1
       request.getRequestDispatcher("welcome.jsp").forward(request,response);
      //页面转向方式 2
        // response.sendRedirect("welcome.jsp");
   } else {
      response.sendRedirect("error.jsp");
   }
%>
</body>
</html>
```

示例 3-5：

```
<%@ page contentType="text/html; charset=UTF-8"%>
<html>
<head>
<title>欢迎 welcome.jsp</title>
</head>
<body>
<h1>欢迎您<%=request.getParameter("userName")%>!</h1>
来自 check.jsp 的信息是:<%=request.getAttribute("msg") %>
</body>
</html>
```

示例 3-6：

```
<%@ page contentType="text/html; charset=UTF-8"%>
<html>
<head>
<title>error.jsp</title>
</head>
<body>
<h3>你的用户名或密码不正确！</h3>
<a herf="index.jsp">回到登录页面</a>
</body>
</html>
```

index.jsp 运行结果如图 3-3 所示。

用户输入正确的用户名和密码后单击“提交”按钮，显示如图 3-4 所示的页面，否则显示如图 3-5 所示的页面。

图 3-3

图 3-4

图 3-5

上面的例子说明了使用 JSP 页面能实现控制作用，充当控制器的角色。实际上页面控制权的转移有两种方式，一种是使用 RequestDispatcher 对象的 forward 方法，即示例 3-4 代码的黑体部分，另一种方法是 response 内置对象的 sendRedirect 方法，即示例 3-4 的注释代码部分。

这两种方式虽然都实现了页面间的转向，但有个小的区别，第一种方式是站点内(同一个 Web 应用内)的页面的跳转，第二种方式是页面的重定向，可以转到站点外。使用第二种方式，放在 request 对象里的数据不能传递给转入的页面，因为是不同的 request 对象。如果把 check.jsp 页面的转向方式改成第二种方式，那么，放在 request 对象中的 msg 属性的值在 welcome.jsp 页面里是取不到的，同时，使用 request.getParameter("userName")也取不到值，这两个值都是 null，如图 3-6 所示。

图 3-6

3.1.3　out 对象

out 对象用于将内容写入 JSP 页面实例的输出流中，它是 javax.servlet.jsp.JspWriter 类的一个实例，out 主要用来控制管理输出的缓冲区(buffer)和输出流(output stream)。常用的方法如表 3-8 所示。

表 3-8

方　法	说　明
void clear()	清除输出缓冲区的内容，但不输出数据到客户端
void clearBuffer()	清除输出缓冲区的内容，并且输出数据到客户端
void close()	关闭输出流，清除所有的内容
int getBufferSize()	返回目前缓冲区的大小(KB)
int getRemaining()	返回目前使用后还剩下的缓冲区大小(KB)
boolean isAutoFlush()	返回缓冲区是否会自动刷新
void flush()	刷新输出缓冲区和输出流

out 对象的两个重要方法为 print()和 println()。当调用 print()方法时，文本被打印到页面上。println()方法类似于 print()方法，在这里 println()方法不会在页面上产生换行的效果，

要换行的话，需要在页面上打印 HTML 的“
”标签。

3.2 作用域通信和控制对象

3.2.1 session 对象

我们知道，当向 Web 服务器请求资源时，Web 服务器会创建一个会话对象来关联这个请求，并把用户的状态保存在会话对象里，实现会话跟踪。JSP 中，使用内置的 session 对象来表示会话对象。session 对象实现了 javax.servlet.http.HttpSession 接口，提供了如表 3-9 所示的常用方法。

表 3-9

方 法	说 明
long getCreationTime()	返回 session 产生的时间，单位是毫秒
String getId()	返回 session 的 id
long getMaxInactiveInterval()	返回最大 session 不活动的时间，若超过这段时间，session 将会失效，单位是秒
long getLastAccessedTime()	返回用户最后通过这个 session 送出请求的时间
void invalidate()	让 session 对象失效
boolean isNew()	判断 session 是否为“新”的
void setMaxInactiveInterval (int interval)	设置 session 最大不活动的时间，若超过这段时间，session 将会失效，单位是秒

HttpSession 接口也提供了如表 3-10 所示的存储和删除属性的方法。

表 3-10

方 法	说 明
void setAttribute(String name, Object value)	把指定的 name 属性和值保存到 session 对象中
Object getAttribute(String name)	根据指定的 name 属性返回保存在 session 对象中的属性值
Enumeration getAttributeNames()	以枚举类型返回 session 对象中所有的属性名

在 JSP 页面里，默认情况下是由容器自动生成与请求相关联的会话对象，并绑定到内置 session 对象。如果在<%@page session="false"%>指令中设置不创建会话对象，那么在页面里使用 session 对象是个错误。我们知道，session 默认的生存时间是 30 分钟，那么在 session 的生存期内，处于同一个 session 的页面可以共享放在 session 对象里的数据，从而实现页

面间的通信。

Session 常用作登录，可以保存已经登录用户的信息，下面的例子演示了使用 Session 完成登录操作。用户只要输入正确的用户名和密码，即可登录成功。登录成功后，我们再次打开首页时无须再次输入账号和密码，当过了对 session 设定的一分钟有效期后，刷新页面，会提示请登录。

新建一个 Dynamic Web Project，名为 JSPWeb2，新建一个类，名为 Login，放在 service 包下，用于执行登录操作。代码如示例 3-7 所示。

示例 3-7：

```
package service;
public class Login {
/**
 * login 方法的作用是比对用户名和密码与预设的是否一致，一致返回 true，不一致返回 false
 */
public boolean Login(String name, String password) {
	if (name.equals("hopu") && password.equals("123456")) {
		return true;
	} else {
		return false;
	}
 }
}
```

在 WebContent 下新建 login.jsp 页面，这个就是用户的登录页面。代码如示例 3-8 所示。

示例 3-8：

```
<%@ page language="java" contentType="text/html; charset=utf-8"
	pageEncoding="utf-8"%>
<!DOCTYPE html PUBLIC "-//W3C//DTD HTML 4.01 Transitional//EN"
"http://www.w3.org/TR/html4/loose.dtd">
<html>
<head>
<meta http-equiv="Content-Type" content="text/html; charset=utf-8">
<title>登录页面</title>
</head>
<body>
	<form action="login_check.jsp" method="post">
		<input type="text" name="user" placeholder="请输入用户名"/>
		 <br/>
		 <br/>
		<input type="password" name="password" placeholder="请输入密码"/>
		<br/>
		<br/>
		<input type="submit" value="提交"/>
	</form>
```

```
</body>
</html>
```

在 WebContent 下新建 login_check.jsp 页面，用来校验用户是否登录成功，并且设置了 session 的存活时间为一分钟。代码如示例 3-9 所示。

示例 3-9：

```
<%@ page language="java" contentType="text/html; charset=UTF-8"
    pageEncoding="UTF-8"%>
    <%@page import="service.*"%>>
<!DOCTYPE html>
<html>
<head>
<meta charset="UTF-8">
<title>登录校验</title>
</head>
<body>
<%
    Login in = new Login();
    String name = request.getParameter("user");
    String pass = request.getParameter("password");
    boolean isLoginSucc = in.Login(name, pass);
    if(isLoginSucc)
    {
    out.println("<script>alert('登录成功！ ');window.location.href='../JSPWeb2/index.jsp'</script>");
    session.setAttribute("user", name);
    session.setMaxInactiveInterval(60);
    }
    else
    {
    out.println("<script>alert('登录失败！ ');window.location.href='../JSPWeb2/login.jsp'</script>");
    }
%>
</body>
</html>
```

在 WebContent 目录下新建 index.jsp 页面，为用户登录成功后所看到的页面，代码如示例 3-10 所示。

示例 3-10：

```
<%@ page language="java" contentType="text/html; charset=utf-8"
    pageEncoding="utf-8"%>
<!DOCTYPE html PUBLIC "-//W3C//DTD HTML 4.01 Transitional//EN"
"http://www.w3.org/TR/html4/loose.dtd">
<html>
<head>
```

```
<meta http-equiv="Content-Type" content="text/html; charset=utf-8">
<title>欢迎</title>
</head>
<body>
  <%
    if(session.getAttribute("user") == null)
    {
     out.println("<script>alert('请先登录');window.location.href='login.jsp'</script>");
     return;
    }
    Object user = session.getAttribute("user");
    out.println("欢迎"+user);
  %>
  <br/>
  <form action="#" method="post">
    <button type="submit" formaction="logout.jsp">登出</button>
  </form>
  </body>
</html>
```

完成以上步骤后，启动项目，在浏览器栏中输入 http://localhost:8080/JSPWeb2，运行结果如图 3-7 所示。

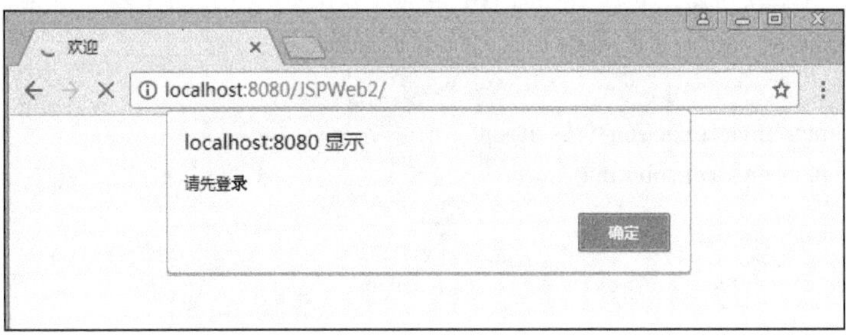

图 3-7

单击“确定”按钮跳到登录界面，输入用户名“hopu”和密码“123456”，跳到登录成功界面，如图 3-8 和图 3-9 所示。

图 3-8

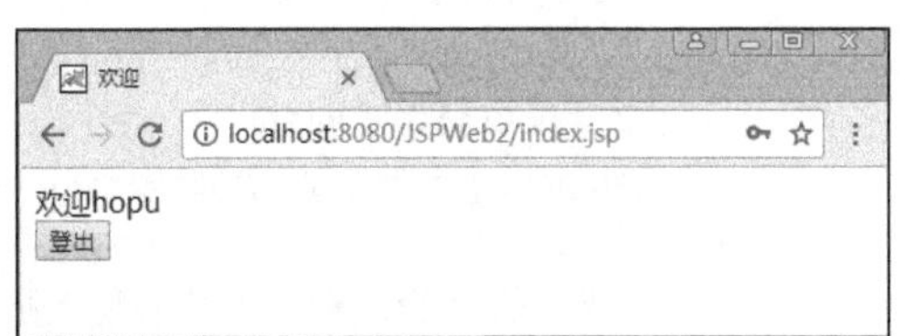

图 3-9

在登录成功后我们刷新页面或者重新打开 index.jsp 页面，依旧会进入“欢迎”页。一分钟后，保存的 session 对象过期，再次刷新首页，会提示请先登录，如图 3-10 所示。

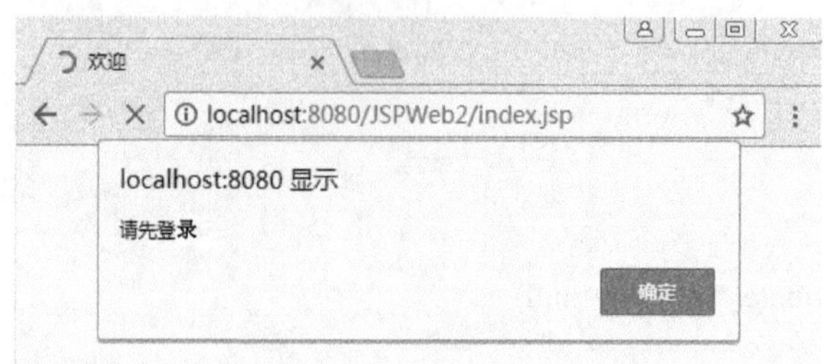

图 3-10

了解 session 对象后，下面我们了解一下 application 对象，它用于获取应用程序上下文环境中的信息。application 对象在容器启动时实例化，在容器关闭时销毁，作用域为整个 Web 容器的生命周期。

下面的例子使用 application 对象实现了投票，不同的用户打开不同的浏览器，都能看到以前的投票结果。vote.jsp 页面是投票页面，doVote.jsp 页面是处理投票的页面，result.jsp 是显示投票结果页面，代码分别如示例 3-11、示例 3-12 和示例 3-13 所示。

示例 3-11：

```
<%@ page contentType="text/html; charset=UTF-8"%>
<html>
    <head><title>新浪新闻调查</title>
    <script type="text/javascript">
     function fun()
     {
          document.myform.action="result.jsp";
          document.myform.submit();
     }
    </script>
    </head>
    <body>
        <h3>新浪新闻调查</h3>
        <form action="doVote.jsp" method="get" name="myform">
            1.您认为次贷危机还需要多长时间才能结束？<br>
            <input type="radio" name="time0" value="1" checked>
            1 年以内<br>
            <input type="radio" name="time0" value="2">
            2 年或者更长时间<br>
            <input type="radio" name="time0" value="3">
            不好说<br>
            2.您对未来全球股市继续恢复上涨是否还有信心？<br>
            <input type="radio" name="have0" value="1" checked>有<br>
            <input type="radio" name="have0" value="2">无<br>
            <input type="radio" name="have0" value="3">不好说<br>
            3.您会否在当前市况下抄底全球资本市场？<br>
            <input type="radio" name="can0" value="1" checked>会<br>
            <input type="radio" name="can0" value="2">不会<br>
```

```
            <input type="radio" name="can0" value="3">不好说<br><br>
            <input type="submit" value="提交">  
            <input type="button" value="查看" onclick="fun()">
        </form>
    </body>
</html>
```

运行结果如图 3-11 所示。

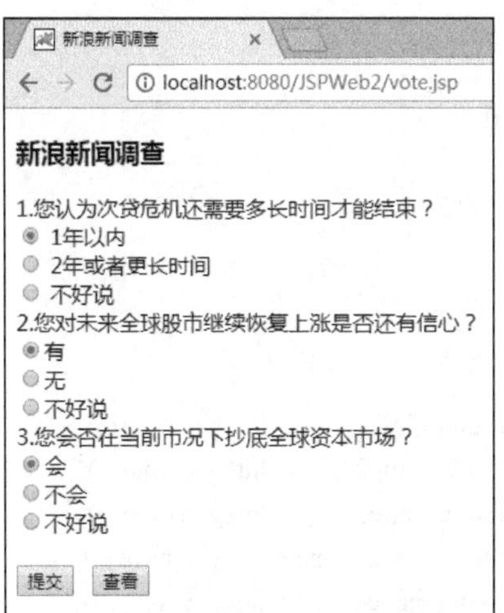

图 3-11

当单击“提交”按钮时，页面转到 doVote.jsp 页面，代码如示例 3-12 所示。该页面只做投票的处理和页面的转向。页面自动转到 result.jsp 页面。

示例 3-12：

```
<%@ page contentType="text/html; charset=UTF-8"%>
<html>
    <head><title>计算投票</title></head>
    <body>
        <%!
        int total = 0 ;
        int time1 = 0, time2 = 0, time3 = 0;
        int have1 = 0, have2 = 0, have3 = 0;
        int can1 = 0, can2 = 0, can3 = 0;
        %>
        <%
        String time0 = request.getParameter("time0");
        String have0 = request.getParameter("have0");
        String can0 = request.getParameter("can0");
        if (time0.equals("1")) {
            time1++;
        } else if (time0.equals("2")) {
            time2++;
```

```
            } else {
                time3++;
            }
            if (have0.equals("1")) {
                have1++;
            } else if (have0.equals("2")) {
                have2++;
            } else {
                have3++;
            }
            if (can0.equals("1")) {
                can1++;
            } else if (can0.equals("2")) {
                can2++;
            } else {
                can3++;
            }
            application.setAttribute("time1", new Integer(time1));
            application.setAttribute("time2", new Integer(time2));
            application.setAttribute("time3", new Integer(time3));
            application.setAttribute("have1", new Integer(have1));
            application.setAttribute("have2", new Integer(have2));
            application.setAttribute("have3", new Integer(have3));
            application.setAttribute("can1", new Integer(can1));
            application.setAttribute("can2", new Integer(can2));
            application.setAttribute("can3", new Integer(can3));
            total ++;
            System.out.println(total);
            application.setAttribute("total",new Integer(total));
            request.getRequestDispatcher("result.jsp").forward(request,
                    response);
            %>
    </body>
</html>
```

result.jsp 页面负责显示投票的结果，代码如示例 3-13 所示。

示例 3-13：

```
<%@ page contentType="text/html; charset=UTF-8"%>
<html>
    <head>
   <title>显示投票结果</title>
   <style type="text/css">
     table {
    border-style: solid;
    border-width: 1px;
}
```

```
</style>
    </head>
    <body>
        次贷危机加剧
        <br>
        <%
            int t1 = ((Integer) application.getAttribute("time1")).intValue();
            int t2 = ((Integer) application.getAttribute("time2")).intValue();
            int t3 = ((Integer) application.getAttribute("time3")).intValue();
            int h1 = ((Integer) application.getAttribute("have1")).intValue();
            int h2 = ((Integer) application.getAttribute("have2")).intValue();
            int h3 = ((Integer) application.getAttribute("have3")).intValue();
            int c1 = ((Integer) application.getAttribute("can1")).intValue();
            int c2 = ((Integer) application.getAttribute("can2")).intValue();
            int c3 = ((Integer) application.getAttribute("can3")).intValue();
            int total = ((Integer) application.getAttribute("total")).intValue();
        %>
        共有<%=total %>人参加
        <br>
        <table border="1" cellpadding="0" cellspacing="0" width="400">
            <tr>
                <td colspan="4">您认为次贷危机还需要多长时间才能结束?</td>
            </tr>
            <tr>
                <td> </td><td>选项</td><td>比例</td><td>票数</td>
            </tr>
            <tr>
                <td>1</td>
                <td>1 年以内</td>
                <td><%=((float)t1)/(t1+t2+t3)*100%>%</td>
                <td><%=t1 %></td>
            </tr>
            <tr>
                <td>2</td>
                <td>2 年或者更长时间</td>
                <td><%=((float)t2)/(t1+t2+t3)*100%>%</td>
                <td><%=t2 %></td>
            </tr>
            <tr>
                <td>3</td>
                <td>不好说</td>
                <td><%=((float)t3)/(t1+t2+t3)*100%>%</td>
                <td><%=t3 %></td>
            </tr>
        </table><br>
        <table border="1" cellpadding="0" cellspacing="0"width="400">
            <tr>
```

```
            <td colspan="4">
             您对未来全球股市继续恢复上涨是否还有信心？
            </td>
        </tr>
        <tr>
            <td> </td><td>选项</td><td>比例</td><td>票数</td>
        </tr>
        <tr>
            <td>1</td>
            <td>有</td>
            <td><%=((float)h1)/(h1+h2+h3)*100%>%</td>
            <td><%=h1 %></td>
        </tr>
        <tr>
            <td>2</td>
            <td>无</td>
            <td><%=((float)h2)/(h1+h2+h3)*100%>%</td>
            <td><%=h2 %></td>
        </tr>
        <tr>
            <td>3</td>
            <td>不好说</td>
            <td><%=((float)h3)/(h1+h2+h3)*100%>%</td>
            <td><%=h3 %></td>
        </tr>
    </table>
    <br>
    <table border="1" cellpadding="0" cellspacing="0"width="400">
        <tr>
            <td colspan="4">您会否在当前市况下抄底全球资本市场？</td>
        </tr>
        <tr>
            <td> </td><td>选项</td><td>比例</td><td>票数</td>
        </tr>
        <tr>
            <td>1</td>
            <td>会</td>
            <td><%=((float)c1)/(c1+c2+c3)*100%>%</td>
            <td><%=c1 %></td>
        </tr>
        <tr>
            <td>2</td>
            <td>不会</td>
            <td><%=((float)c2)/(c1+c2+c3)*100%>%</td>
            <td><%=c2 %></td>
        </tr>
        <tr>
```

```
                <td>3</td>
                <td>不好说</td>
                <td><%=((float)c3)/(c1+c2+c3)*100%>%</td>
                <td><%=c3 %></td>
            </tr>
        </table>
    </body>
</html>
```

result.jsp 页面运行结果如图 3-12 所示。

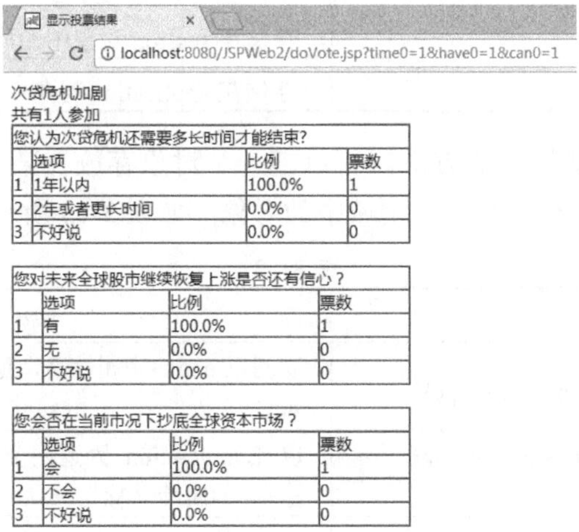

显示投票结果

localhost:8080/JSPWeb2/doVote.jsp?time0=1&have0=1&can0=1

次贷危机加剧

共有1人参加

您认为次贷危机还需要多长时间才能结束?			
	选项	比例	票数
1	1年以内	100.0%	1
2	2年或者更长时间	0.0%	0
3	不好说	0.0%	0

您对未来全球股市继续恢复上涨是否还有信心？			
	选项	比例	票数
1	有	100.0%	1
2	无	0.0%	0
3	不好说	0.0%	0

您会否在当前市况下抄底全球资本市场？			
	选项	比例	票数
1	会	100.0%	1
2	不会	0.0%	0
3	不好说	0.0%	0

图 3-12

3.2.2　pageContext 对象

pageContext 对象是 javax.servlet.jsp.PageContext 类的实例，容器自动为每个 JSP 页面创建与之对应的 pageContext 对象，并把与 JSP 对应的 Servlet 对象的相关对象都自动地加入到 pageContext 对象中，所以，用户可以使用 pageContext 对象来获得与 JSP 页面对应的 Servlet 对象的相关对象。使用 pageContext 对象来获取 Servlet 对象的组件的方法如表 3-11 所示。

表 3-11

方　法	说　明
Exception getException()	返回当前网页的异常，不过此网页必须是错误页
JspWriter getOut()	返回当前网页的输出流 out 对象
Object getPage()	返回当前网页的 Servlet 实例 page 对象
ServletRequest getRequest()	返回当前网页的请求 request 对象
ServletResponse getResponse()	返回当前网页的响应 response 对象
ServletConfig getServletConfig()	返回当前网页的 config 对象
ServletContext getServletContext()	返回当前网页的 context 对象
HttpSession getSession()	返回与当前网页关联的 session 对象

在前面我们学习过，可以把一个对象存放在request对象中或者是session对象中和application对象中，那么这个对象便有了不同的作用域。当然，我们也可以把一个对象存放在pageContext中，那么这个对象便有了页面作用域。pageContext对象提供了 4 个属性来分别对应这 4 个作用域，如表 3-12 所示。

表 3-12

属　性	说　明
PAGE_SCOPE	存储在 pageContext 对象中的属性的作用域
REQUEST_SCOPE	存储在 request 对象中的属性的作用域
SESSION_SCOPE	存储在 session 对象中的属性的作用域
APPLCATION_SCOPE	存储在 application 对象中的属性的作用域

使用pageContext对象提供的方法，可以把一个对象存放在某个作用域，也可以使用pageContext提供的方法来获取某一作用域中的对象，如表 3-13 所示。

表 3-13

方　法	说　明
Object getAttribute(String name,int scope)	返回以 name 为属性名，范围为 scope 的指定的属性对象
Enumeration getAttributeNamesInScope(int scope)	以 Enumeration 为返回类型，返回所有属性范围为 scope 的属性名称
int getAttributesScope(String name)	返回属性名称为 name 的属性范围
void removeAttribute(String name)	移除属性名称为 name 的属性对象
void removeAttribute(String name, int scope)	移除属性名称为 name、范围为 scope 的属性对象
void setAttribute(String name, Object value, int scope)	设定属性对象的名称为 name、值为 value、范围为 scope
Object findAttribute(String name)	寻找在所有范围中属性名称为 name 的对象

下面的代码片段演示了 pageContext 的用法，分别是把属性存储到 pageContext、request、session 和 application 对象中，与注释部分的代码的作用是相同的。

```
<%
  pageContext.setAttribute("kongfu","太极拳");
  //pageContext.setAttribute("kongfu","太极拳",pageContext.PAGE_SCOPE);

  request.setAttribute("name","张无忌");
  //pageContext.setAttribute("name","张无忌",pageContext.REQUEST_SCOPE);

  session.setAttribute("age","108");
  //pageContext.setAttribute("age","108",pageContext.SESSION_SCOPE);

  application.setAttribute("address","湖北武当山");
  //pageContext.setAttribute("address","湖北武当山",pageContext.APPLICATION_SCOPE);
  %>
```

3.3 Servlet 相关对象

3.3.1 page 对象

page 对象代表 JSP 本身，更准确地说它表示与 JSP 对应的 Servlet 类的实例。page 指令的 extends 属性用于显示指定 Servlet 类，否则在构造 Servlet 时，JSP 容器将使用特定的类。在任何一种情况下，都需要有 Servlet 类来实现 javax.servlet.jsp.JspPage 接口。

3.3.2 config 对象

config 对象实现了 javax.servlet.ServletConfig 接口，用来存放与 JSP 相关的初始化数据。ServletConfig 的方法如表 3-14 所示。

表 3-14

方 法	说 明
String getInitParameter(name)	返回指定初始化参数的值
Enumeration getInitParameterNames()	返回所有初始化参数的名称
ServletContext getServletContext()	返回 Web 应用的 servletContext
String getServletName()	返回在 Web 应用发布描述器(web.xml)中指定的 servlet 的名字

我们来看一个例子，在 showconfig.jsp 页面中，使用 config 对象来从 web.xml 读取 JSP 页面的初始化参数。代码如示例 3-14 所示。

示例 3-14：

```
<%@ page contentType="text/html; charset=UTF-8"%>
<html>
  <head>
    <title>config 的用法</title>
  </head>
  <body>
    下面的参数来自 web.xml 配置文件：<br>
    姓名:<%=config.getInitParameter("name") %><br>
    派别:<%=config.getInitParameter("pai") %><br>
    地方:<%=config.getInitParameter("location")%><br>
  </body>
</html>
```

我们必须在 web.xml 中使用<servlet>标签来配置初始化参数。与 Servlet 的初始化参数的配置是一样的，这里要注意使用<jsp-file>标签，而不像 Servlet 使用的是<servlet-class>

标签，如示例 3-15 所示。

示例 3-15：

```
<servlet>
<servlet-name>show</servlet-name>
<jsp-file>/showconfig.jsp</jsp-file>
<init-param>
<param-name>name</param-name>
<param-value>张三丰</param-value>
</init-param>
<init-param>
<param-name>pai</param-name>
<param-value>武当派</param-value>
</init-param>
<init-param>
<param-name>location</param-name>
<param-value>湖北.武当山</param-value>
</init-param>
</servlet>
<servlet-mapping>
<servlet-name>show</servlet-name>
<url-pattern>/show.jsp</url-pattern>
</servlet-mapping>
```

需要注意的是，此时用户请求的 URL 地址不再是 showconfig.jsp，而是 show.jsp，除非把<url-pattern>配置成 showconfig.jsp。示例 3-14 的运行结果如图 3-13 所示。

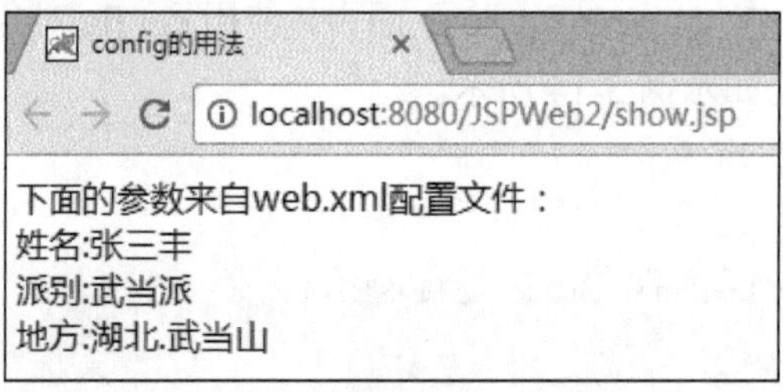

图 3-13

3.4 错误处理对象

JSP 错误处理涉及一个对象，即 exception 对象，此隐式对象只适用于 JSP 错误页面。在某个 JSP 页面抛出异常时，将转发至 JSP 错误处理页面。exception 内置对象是 java.lang.Throwable 类的实例，提供此对象是为了在 JSP 中处理异常。若要在页面中使用 exception 对象，必须在 page 指令中指定<%@ page isErrorPage="true" %>。Throwable 类提供了如表 3-15 所示的方法。

表 3-15

方　法	说　明
String getMessage()	返回与此异常关联的描述性错误信息
void printStackTrace(out)	向指定的输出流输出堆栈跟踪
String toString()	返回一个包括异常名及错误消息的字符串

示例 3-16 和示例 3-17 所示为一个引发异常的页面 math.jsp，当发生异常的时候将显示一个异常页面 error.jsp。

示例 3-16：

```
<%@ page errorPage="error.jsp" contentType="text/html; charset=UTF-8"%>
<html>
<body>
表达式 5/0 的值是：
<%=5/0%>
</body>
</html>
```

示例 3-17：

```
<%@ page contentType="text/html; charset=UTF-8" isErrorPage="true"%>
<html>
<body>
<h3>发生了异常!异常信息如下:<br>
<%= exception.getMessage()%>
</h3></body>
</html>
```

运行结果如图 3-14 所示。

发生了异常!异常信息如下:
/ by zero

图 3-14

当 math.jsp 页面被装载时会产生一个除数为零的异常，而且指定了错误处理页面 error.jsp。所以访问 math.jsp 时会直接转到 error.jsp 页面。

【单元小结】

- JSP 隐式对象可分为 4 个类别：输入/输出对象、作用域通信和控制对象、Servlet 相关对象、错误处理对象。
- request 对象封装了请求信息，response 对象封装了输出结果。
- out 对象表示与响应关联的实际输出流。
- session 隐式对象主要用于存储和检索属性值，作用于会话中。
- application 隐式对象用来存储运行环境的有关信息，它能被所有 JSP 和 Servlet 实例访问，且从服务器启动时被创建直到服务器停止。
- pageContext 隐式对象表示 JSP 本身的运行环境，提供对所有其他隐式对象及其属性的访问。
- exception 隐式对象用于在 JSP 中处理异常。

【单元自测】

1. 如果某一 JSP 页面的表单中，有几个复选框，name 都为“habit”，则该 JSP 提交后，通过下面(　　)语句去获取用户选中复选框的值。

   A. request.getAttribute("habit");

   B. request.getParameter("habit");

   C. request.getParameterValues("habit");

   D. request.getHabit();

2. application 的(　　)方法返回 Servlet 容器支持的 Servlet API 的主要版本。

   A. getServerInfo()　　　　B. getMajorVersion()

   C. getMinorVersion()　　　D. 以上都不是

3. exception 内置对象的(　　)方法返回与此异常关联的描述性错误信息。

   A. getInformation()　　　B. getMessage()

   C. getException()　　　　D. 以上都不是

4. JSP 中的(　　)隐式对象表示 Servlet 类的实例。

   A. request　　　B. response

   C. page　　　　 D. session

【上机实战】

上机目标

- 掌握 request 对象的使用。
- 掌握 response 对象的使用。

- 掌握 session 对象的使用。
- 掌握 exception 对象的使用。

上机练习

◆ 第一阶段 ◆

练习 1：request 和 response 对象的使用

【问题描述】

管理员登录并显示“欢迎管理员”。

【问题分析】

用户在登录页 login.html 页面输入用户名和密码，提交到 admin.jsp 页面，判断用户名和密码是否正确。如果正确则显示“欢迎管理员”，否则返回到 login.html 页面。

【参考步骤】

(1) 创建 login.html 文件，代码如示例 3-18 所示。

示例 3-18：

```
<%@ page contentType="text/html; charset=UTF-8"%>
<html>
<head><title>登录</title></head>
<body>
<form name="form1" method="post" action="admin.jsp">
用户名:<input name="userName" type="text" ><br>
密     码:<input name="password" type="password" value=""><br>
<input type="submit" value="提交">
</form></body>
</html>
```

(2) 创建 admin.jsp 文件，代码如示例 3-19 所示。

示例 3-19：

```
<%@ page contentType="text/html; charset=UTF-8"%>
<html>
    <head>
        <title>欢迎管理员</title>
    </head>
    <body>
        <%
            if (request.getParameter("userName").equals("admin")
```

```
                        && request.getParameter("password").equals("admin")) {
        %>
        <h1>
            欢迎您管理员!
        </h1>
        <%
            } else {
                response.sendRedirect("login.jsp");
            }
        %>
    </body>
</html>
```

(3) 运行结果如图 3-15 和图 3-16 所示。

图 3-15

图 3-16

练习 2：session 对象的使用

【问题描述】

显示当前用户访问某个网页的次数。

【问题分析】

(1) 使用 session 对象来存储和检索属性值，声明一个 String 类型的变量 count，使用 session 的 getAttribute()方法得到属性值，并把该值赋给 count。通过一个方法将 count 的值加 1，并存储在 session 对象中。

(2) 注意，第一次访问页面的时候 count 得到的是一个 null 值，所以要做一个判断，如果 count 的值为 null，就赋值为 1。

【参考步骤】

创建 sessionDemo.jsp 文件，代码如示例 3-20 所示。

示例 3-20：

```
<%@ page contentType="text/html; charset=UTF-8"%>
<html>
<body><h1>
<%
    String count = (String)session.getAttribute("numVisits");
    count = addCount(count);
```

```
    out.print("您访问了本页面" + count + "次!");
    session.setAttribute("numVisits",count);
%>
<%!
    String addCount(String count){
        if(count!=null)
            return Integer.toString(Integer.parseInt(count)+1);
        else
            return "1";
    }
%>
</h1></body>
</html>
```

运行结果如图 3-17 所示。

您访问了本页面6次!

图 3-17

◆ 第二阶段 ◆

练习 3：exception 对象的使用

【问题描述】

对产生数组下标越界错误的页面创建一个相应的错误页面。

【问题分析】

(1) 创建一个引发错误的页面 makeError.jsp。

(2) 在 makeError.jsp 中使用 JSP 脚本元素代码段引发一个数组下标越界的异常，并使用 page 指令指定错误产生后所转发的页面。

(3) 创建错误页面 error.jsp，显示错误信息。

(4) 一定要在 error.jsp 中写上<%@page isErrorPage=" true ">，告诉 JSP 这是一个错误处理页面。

【拓展作业】

1. 编写一个程序，显示 request 的 HTTP 方法、用户的 IP 地址和主机名。
2. 编写一个程序，获得会话的 id、最大失效时间和最后访问的时间。

3. 编写一个 JSP 页面，用来设置 application 对象的属性值，另一个 JSP 页面用来获得这个属性值。

4. 编写一个 JSP 页面，用户输入两个数，计算并输出结果。如果产生了错误，则在另一个页面中输出错误的信息。

单元四 Servlet 入门

课程目标

- 掌握如何编写 Servlet 类
- 了解 Servlet 的生命周期
- 使用 Servlet 进行数据库操作

简介

Servlet是Java 2.0中新增的一个全新功能，Servlet是采用Java技术来实现CGI功能的一种技术。Servlet与CGI一样都运行在Web服务器上，用来生成Web页面。对Servlet来说，在Java虚拟机上，每一个请求由一个“小”Java线程(thread)响应，而不是一个“大”操作系统进程。这样则若干个请求对于Servlet来说，可产生若干个线程，但只有一个Servlet代码的拷贝。

Servlet是与平台无关的服务器组件，使用Java编写，遵循标准API。它能直接或借助插件在几乎所有的Web服务器上运行。其中，服务器负责Servlet和客户端的通信，采用“请求/响应”模式。Servlet能自动粘贴和解码HTML表单数据，读取和设置HTTP头，处理Cookie，跟踪Session等其他功能。Servlet之间能共享数据，很容易实现诸如数据库连接池的功能，更能方便地实现会话管理和Servlet之间的通信。

4.1 Servlet API

Servlet的核心是javax.servlet.Servlet接口，所有的Servlet必须实现这个接口。我们写的Servlet大都是从GenericServlet或HttpServlet类进行扩展来实现的。当扩展了采用HTTP通信协议的HttpServlet，并把这个Servlet动态地加载到Servlet服务器上时，就能够使用HTTP请求和HTTP响应与客户端进行交互。Servlet容器支持请求和响应所用的HTTP协议如图4-1所示。

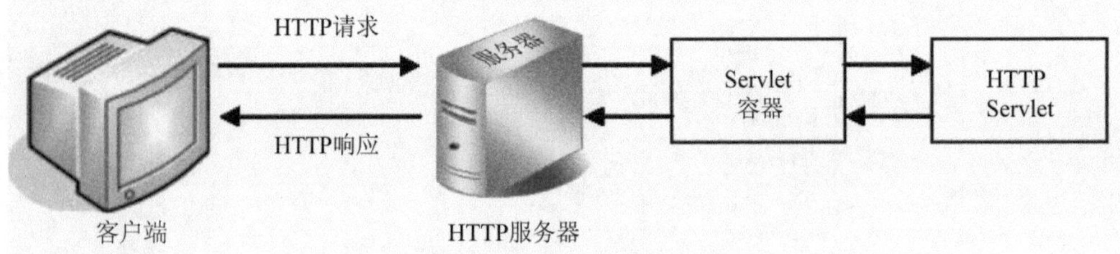

图4-1

Servlet API包含于两个包中，分别是javax.servlet和javax.servlet.http。常用的接口和类之间的关系如图4-2所示。更多的类和接口参看Java EE API。

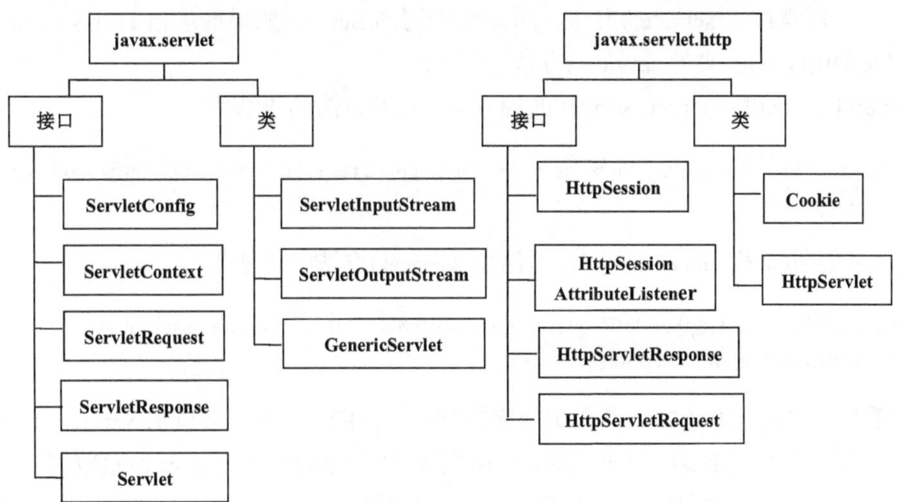

图 4-2

4.2 扩展 Servlet

从图4-2可以看出，编写Servlet既可以实现javax.servlet.Servlet接口，也可以从javax.servlet.GenericServlet或javax.servlet.http.HttpServlet实现类来继承。其中，GenericServlet实现了Servlet接口，而 HttpServlet 又继承了 GenericServlet。所以，我们编写 Servlet 的时候，一般可以从实现类来继承其中的一个。

4.2.1 继承 GenericServlet

GenericServlet 是个抽象类，其中的 service()方法定义为抽象方法。所以，如果我们写的 Servlet 从 GenericServlet 来继承的话，就必须要实现该抽象类中定义的 service()方法。service()方法的语法为：

```
public abstract void service(ServletRequest request, ServletResponse response) throws ServletException,
    IOException
```

Service()方法有两个入参：ServletRequest 和 ServletResponse。Servlet 容器将客户请求的信息封装到 ServletRequest 对象中，并传给 service()方法，经过处理后，把响应的结果封装到 ServletResponse 对象中传回给客户端。而这种方式并不常用，更多的还是采取下面介绍的继承 HttpServlet 方式。

4.2.2 继承 HttpServlet

HttpServlet 类扩展了 GenericServlet 类，它不仅重写了 GenericServlet 类中的 service 抽

象方法，而且还重载了 service()方法，所以我们写的 Servlet 如果继承自 HttpServlet，通常不必再重写 HttpServlet 类的 service()方法。

HttpServlet 中重写 GenericServlet 的抽象 service()方法的语法为：

```
public void service(ServletRequest request, ServletResponse response) throws ServletException,
    IOException
```

HttpServlet 中重载 GenericServlet 的抽象 service()方法的语法为：

```
protected void service(HttpServletRequest request, HttpServletResponse response) throws
    ServletException, IOException
```

可以看出，这两个方法的入参和访问修饰符是不同的。当客户端请求 Servlet 时，Servlet 容器将客户请求的信息封装到 ServletRequest 对象中，先调用重写的 service()方法，在该方法中把 ServletRequest 对象和 ServletResponse 对象强制转换成 HttpServletRequest 对象和 HttpServletResponse 对象，并把这两个转换后的对象作为入参，传给重载的 service()方法。在该方法中，先从 HttpServletRequest 对象中获取 HTTP 请求的方式，然后根据请求方式调用相对应的方法。HTTP 的请求方式包括 DELETE、GET、OPTIONS、POST、PUT、TRACE，所以在 HttpServlet 中加入了与这些请求方式相对应的方法，这些方法可以被处理 HTTP 请求的 service 方法自动地调用。这些方法是 doHead()、doGet()、doPost()、doDelete()、doOptions()、doPut()、doTrace()。

- doHead()：用来处理 HTTP 客户端的 HEAD 请求。如果客户端只希望查看响应头中的部分信息，如 Content-Type 和 Content-Length，可以发送 HEAD 请求。如果重写此方法，必须避免处理响应体，而直接设置响应头用来提高性能。如果请求的格式不正确，方法 doHead()将返回一个 HTTP 的 Bad_Request 信息。doHead()方法的语法为：

```
public void doHead (HttpServletResquest request,HttpServletResponse response) throws ServletException,
    IOException
```

- doGet()：如果 HTTP 客户端发送给服务器的请求是 GET 方式，这个方法将被 service()方法调用。GET 操作仅仅允许客户从 HTTP Server 上“取得”资源。重写此方法不仅支持 HTTP GET 请求，而且还自动支持 HTTP HEAD 请求。一个 HTTP HEAD 请求其实是没有响应体只有响应头的 HTTP GET 请求。当我们重写该方法时，不仅要读取请求数据，还要写响应头，获取响应打印机或输出流对象来写响应数据。如果请求的格式不正确，方法 doGet()将返回一个 HTTP 的 Bad_Request 错误信息。doGet()方法的语法为：

```
public void doGet(HttpServletResquest request,    HttpServletResponse response)    throws ServletException,
    IOException
```

- doPost()：如果HTTP客户端发给服务器的请求是POST方式，这个方法会被service()方法调用。HTTP POST方式允许用户一次向服务器发送不限长度的数据和比较敏感的数据，如卡号和密码等。如果HTTP POST请求的格式不正确，方法doPost()的默认执行将返回一个HTTP的Bad_Request错误信息。doPost()方法的语法为：

```
public void doPost(HttpServletResquest request,HttpServletResponse response) throws
    ServletException,IOException
```

- doDelete()：用来处理HTTP客户端的DELETE请求。此操作允许客户端请求从服务器中删除一个文档或页面。如果HTTP DELETE请求的格式不正确，方法doDelete()的默认执行将返回一个HTTP的Bad_Request错误信息。doDelete()方法的语法为：

```
public void doDelete (HttpServletResquest request,HttpServletResponse response) throws
    ServletException,IOException
```

- doOptions()：用来处理HTTP的OPTIONS()请求。此操作自动决定支持哪些HTTP方式。例如，如果一个Servlet重写doGet()方法，doOptions()方法将会返回如下的信息：Allow、GET、HEAD、TRACE、OPTIONS。一般情况下不需要重载方法doOptions()，除非一个Servlet实现一个新的HTTP方法。doOptions()方法的语法为：

```
public void doOptions (HttpServletResquest request, HttpServletResponse response) throws
    ServletException, IOException
```

此外，还有doPut()方法和doTrace()方法，这里就不一一列举了，有兴趣的读者可以参看Servlet的源代码。

4.2.3　ServletRequest 和 ServletResponse 接口

GenericServlet和HttpServlet的service()方法的两个入参分别是ServletRequest和ServletResponse类型，其实，它们都是接口。ServletRequest接口中封装了客户请求的数据信息，如客户的请求方式、参数名和参数值、客户端正在使用的协议以及发出客户请求的远程主机信息等。ServletRequest接口还为Servlet提供了直接以二进制方式读取客户请求数据流的ServletInputStream。该接口中定义的方法如表4-1所示。

表4-1

方法名	作　用
Object getAttribute(String name)	根据name属性名返回属性值
String getContentType()	返回客户请求数据的MIME类型
ServletInputStream getInputSteam()	返回以二进制方式直接读取客户请求数据的输入流
String getParameter(String name)	根据给定的参数名返回参数值

(续表)

方法名	作　用
String getRemoteAddr()	返回远程客户主机的 IP 地址
String getRemoteHost()	返回远程客户主机的主机名
int getRemotePort()	返回远程客户主机的端口
void setAttribute(String name , Object object)	设置 name 属性名和值

ServletResponse 接口为 Servlet 提供了返回响应结果的方法。它允许 Servlet 设置返回数据的长度和 MIME 类型，并且提供输出流 ServletOutputSteam。

ServletResponse 接口中定义的方法如表 4-2 所示。

表 4-2

方法名	作　用
ServletOutputStream getOutputSteam()	返回可以向客户端发送二进制数据的输出流对象
String getContentType()	返回 Servlet 发送的响应数据的 MIME 类型
PrintWriter getWriter()	返回可以向客户端发送字符数据的 PrinterWriter 对象
String getCharacterEncoding()	返回 Servlet 发送的响应数据的字符编码
void setCharacterEncoding (String charset)	设置 Servlet 发送的响应数据的字符编码
void setContentType(String type)	设置 Servlet 发送的响应数据的 MIME 类型

4.3　Servlet 的生命周期

与人一样，所谓生命周期，实际上是从出生到死亡的过程。Servlet 的生命周期是指 Servlet 实例在被支持 Servlet 的容器创建并装载到 Servlet 容器中，继而提供服务，再到该实例被容器销毁或重新载入的全部过程。与前面实例对象的创建不同的是，我们并没有显式地使用 new 关键字来构造这个 Servlet 实例对象，而这个实例对象是由 Servlet 容器来控制，是容器决定着 Servlet 实例对象的创建和销毁。我们说，当客户端第一次向服务器请求 Servlet 时，容器便创建并载入一个 Servlet 实例，并让这个实例提供一次服务，之后，这个 Servlet 的实例一直存在于服务器的容器里，并等待再次提供服务。如果服务器停止服务或是重新启动服务，那么，这个 Servlet 的实例便被销毁了。在 javax.servlet.Servlet 接口中定义了 3 个方法来描述 Servlet 的生命周期。

4.3.1　Servlet 初始化时期

在下列情形时，Servlet 容器会创建并装载 Servlet。

- 当客户端第一次向服务器请求 Servlet。
- Servlet 容器启动时自动装载某些 Servlet。
- Servlet 的源文件被更新后，重新装载 Servlet。

Servlet 容器会创建一个 Servlet 的实例，并且调用 Servlet 的 init()方法对其进行初始化。在整个 Servlet 的生命周期中，init()方法只会被调用一次。init()方法的语法有以下两种形式：

```
public void init() throws ServletException
```

```
public void init(ServletConfig config) throws ServletException
```

在 Servlet 初始化之前，Servlet 容器会为要初始化的 Servlet 创建一个 ServletConfig 实例对象，把一些初始化的参数封装到该对象中。如果用户在写的 Servlet 中重写了第二种形式(有参数)的 init()方法，则必须先调用 super.init(config)方法来确保参数 config 引用 ServletConfig 对象。如果用户在写的 Servlet 中重写了第一种形式(无参数)的 init()方法，可以不调用 super.init()方法，要在 init()方法中使用 ServletConfig 对象，可以使用 Servlet 类的 getServletConfig()方法获得。

4.3.2 Servlet 响应客户请求时期

在服务器装载并初始化 Servlet 后，Servlet 就能用 service()方法来处理客户端的请求。Servlet 容器创建 ServletRequest 对象和 ServletResponse 对象，并把客户的请求信息封装到 ServletRequest 对象中，service()方法从 ServletRequest 对象中获取请求信息并处理该请求，并通过 ServletResponse 对象向客户发回相应的响应结果。

4.3.3 Servlet 结束时期

Servlet 一直运行到它们被服务器卸载。当 Web 应用被终止、服务器停止服务或是 Servlet 容器重新装载 Servlet 新实例时，Servlet 容器会先调用 Servlet 的 destroy()方法来释放 Servlet 使用的资源。destroy()方法的语法为：

```
public void destroy()   { //回收在 init()中启用的资源，如关闭数据库的连接等。}
```

4.4 HTTP 协议和 HttpServlet

Web 服务器和客户浏览器可以通过 HTTP 协议在 Internet 上发送和接收消息。我们知道，HTTP 协议的特点是一种基于请求/响应的模式，客户端发送一个请求，服务器返回对该请求的响应。并且 HTTP 协议使用可靠的 TCP 连接，默认的端口是 80，最新的版本是 1.1。在 HTTP 中，服务器是被动的，客户端是主动的。客户端要请求服务器的资源，首先要建立与服务器之间的连接，服务器不会主动联系客户端或要求客户端建立连接。一旦服务器传回客户端请求的资源，服务器会断开建立的连接。

4.4.1 HTTP 请求和 HTTP 响应

一个典型的 HTTP 请求由 3 部分组成，分别是请求方法 URI 协议/版本、请求头、请求正文。我们来看一个 HTTP 请求的例子：

```
//1.URI 协议/版本
GET/index.jsp HTTP1.1
//2.请求头
Accept:image/gif,image/jpeg,*.*
Accept-Language:zh-cn
Connection:Keep-Alive
Host: localhost:8080
User-Agent: Mozilla/5.0 (Windows NT 6.1; Win64; x64) AppleWebKit/537.36 (KHTML, like Gecko) Chrome/64.0.3282.186 Safari/537.36
//3.请求正文
name=bush&pwd=123456
```

例子中的第一行是请求的 URI 协议/版本，使用的请求方式是“GET”，请求的 URI 地址是“/index.jsp”，使用的 HTTP 协议版本是“1.1”。第二部分是请求头，包含与客户端的环境和请求正文的相关信息。第三部分是请求正文，包含两个键值对，实际应用中可以包含更多的内容。

HTTP 响应也由 3 部分组成，分别是协议状态代码描述、响应头、响应正文。我们来看一个 HTTP 响应的例子：

```
//1.响应的协议、状态码和描述
HTTP1.1    200    OK
//2.响应头
Server:Apache Tomcat/8.5.32
Date: Wed, 22 Aug 2018 02:02:32 GMT
Content-Type:text/html
Last-Modified: Wed, 22 Aug 2018 02:02:32 GMT
Content-Length:200
//3.响应正文
name=bush&pwd=123456
```

第一行是协议版本，200 表示服务状态码，OK 表示服务器已经成功地处理了客户发出的请求。第二部分包含服务器端的有用信息，如服务器的类型、日期时间、内容类型和长度。第三部分是服务器返回的 HTML 页面。

4.4.2 HttpServletRequest 和 HttpServletResponse 接口

HttpServlet 的重载的 service() 方法的两个入参分别是 HttpServletRequest 和

HttpServletResponse 类型，它们都是接口。前面我们学习过，Servlet 容器在 ServletRequest 接口中封装了客户请求的数据信息，如客户的请求方式、参数名和参数值、客户端正在使用的协议以及发出客户请求的远程主机信息等。在 HttpServlet 重写的 service()方法中，ServletRequest 被强制转换成 HttpServletRequest 类型，并作为参数传给 HttpServlet 重载的 service()方法。HttpServletRequest 接口中定义了读取 HTTP 请求信息的常用方法，如表 4-3 所示。

表 4-3

方法名	作　用
Cookie [] getCookies()	返回 HTTP 请求的 Cookies
String getHeader(String name)	返回参数指定的 HTTP 请求的 Header 数据
String getRequestURI()	返回 HTTP 请求的 URI
String getQueryString()	返回 HTTP 请求数据中的查询字符串
String getMethod()	返回 HTTP 请求方法

前面学习过，ServletResponse 接口为 Servlet 提供了返回响应结果的方法。它允许 Servlet 设置返回数据的长度和 MIME 类型，并且提供输出流 ServletOutputSteam。在 HttpServlet 重写的 service()方法中，ServletResponse 被强制转换成 HttpServletResponse 类型，并作为参数传给 HttpServlet 重载的 service()方法。HttpServletResponse 接口中定义了提供生成响应数据 Header 的方法，如表 4-4 所示。

表 4-4

方法名	作　用
void addCookie(Cookie cookie)	向 HTTP 响应中加入 Cookie
void setHeader(String name , String value)	设置 HTTP 响应的 Header，如果参数 name 对应的 Header 已经存在，则覆盖原来的 Header 数据
void addHeader(String name , String value)	向 HTTP 响应中加入 Header

4.5 Servlet Web 应用开发

4.5.1 Servlet 启动时加载

在前面我们了解到创建 Servlet 实例是第一次被访问的时候，init()方法才会执行。假如在 init()方法中执行了一些耗时操作，如加载一些配置文件并解析可能需要花费数秒钟。如何使第一个用户在第一次访问时无须花费这个时间，就是我们这次要讲解到的启动时加载。

我们通过配置将 Servlet 的实例化过程放在服务器启动的时候，这样实例化所耗费的时间就会在服务器启动的时候一起消耗掉。这就需要配置 load-on-startup，有两种方式，一种是在 web.xml 文件中进行配置，如示例 4-1 所示，另一就是通过@WebServlet (loadOnStartup=1)注解方式，在以后的课程中我们再去学习注解。

示例 4-1：

```
<servlet>
    <!-- Servlet 名 -->
    <servlet-name>TestServlet</servlet-name>
    <!-- Servlet 的实现类 -->
    <servlet-class>control.GetInfoFromDB</servlet-class>
    <!-- 配置应用启动时，创建 Servlet 实例 -->
    <load-on-startup>1</load-on-startup>
</servlet>
```

示例 4-1 中加粗部分数值必须是整数，只有当数值大于或等于 0 时，才会启动时就加载。数值的大小表示的是 Servlet 被载入的优先级，当数值越小时，优先级越高。

4.5.2 Servlet 访问路径配置

在单元三中的示例 3-15 中，我们就已经用到了访问路径的配置，也就是 urlPatterns。该属性就是指定 Servlet 处理的 URL，有以下 3 种路径匹配方式。

1. 完全路径匹配

```
<servlet-mapping>
<servlet-name> MyServlet</servlet-name>
<url-pattern>/show.jsp</url-pattern>
<url-pattern>/hopu/demo.html</url-pattern>
<url-pattern>/hello</url-pattern>
</servlet-mapping>
```

完全路径匹配，就是以/开始，再加上访问文件名。当在浏览器中访问如下几种 url 时，都会匹配到该 Servlet。

```
http://localhost:8080/JSPDemo/show.jsp
http://localhost:8080/JSPDemo/hopu/demo.html
http://localhost:8080/JSPDemo/hello
```

2. 目录匹配

```
<servlet-mapping>
<servlet-name> MyServlet</servlet-name>
<url-pattern>/hopu/*</url-pattern>
</servlet-mapping>
```

目录匹配，就是以/开始，以/*结束。例如/hopu/*，下面几种 url 都能匹配到该 Servlet。

```
http://localhost:8080/JSPDemo/hopu/test.jsp
http://localhost:8080/JSPDemo/hopu/demo.html
http://localhost:8080/JSPDemo/hopu/action
```

3. 扩展名匹配

```
<servlet-mapping>
<servlet-name>MyServlet</servlet-name>
<url-pattern>*.jsp</url-pattern>
<url-pattern>*.action</url-pattern>
<url-pattern>*.do</url-pattern>
</servlet-mapping>
```

扩展名匹配是不允许以/开始的，要以*开始，再加上扩展名。例如，当在浏览器中访问如下几种 url 时，都会匹配到该 Servlet。

```
http://localhost:8080/JSPDemo/test.jsp
http://localhost:8080/JSPDemo/login.action
http://localhost:8080/JSPDemo/check.do
```

4.5.3 使用 Eclipse 开发第一个 Servlet

单元一中我们了解了 Web 应用程序的开发步骤和基本概念，在本例中，使用 Eclipse 开发环境来创建第一个 Servlet，并演示 Servlet 的生命周期。具体步骤如下。

(1) 打开 Eclipse 开发环境，新建一个 Dynamic Web Project，工程名叫作 MyFirstWeb，存放在默认的文件夹中，如图 4-3 所示，单击“Next”按钮，然后在新的界面窗口中再次单击“Next”按钮，弹出如图 4-4 所示的界面，选中“Generate web.xml deployment descriptor”前的复选框。

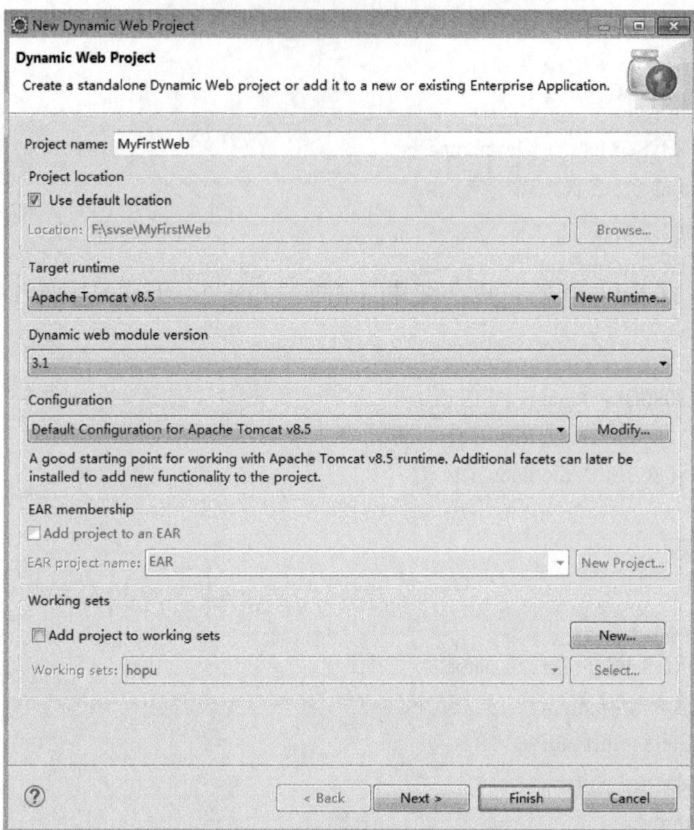

图 4-3

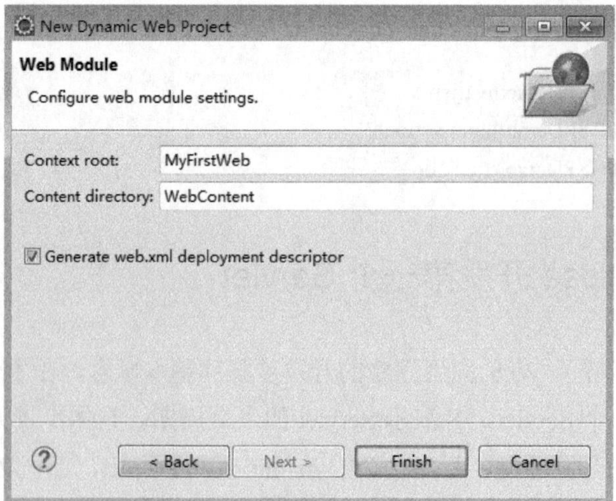

图 4-4

(2) 单击“Finish”按钮，完成 Web 应用工程的创建，然后选择“File→New Servlet”命令，选择 Servlet 菜单，打开创建 Servlet 的界面，如图 4-5 所示。在 Java package 栏输入包名字，Class name 栏输入 Servlet 的名字。

图 4-5

(3) 输入包名和 Servlet 名字后，单击“Next”按钮，新的界面保持默认状态，再次单击“Next”按钮，选中“init”和“destroy”前的复选框，取消选中“doPost”前的复选框，最终结果如图 4-6 所示。

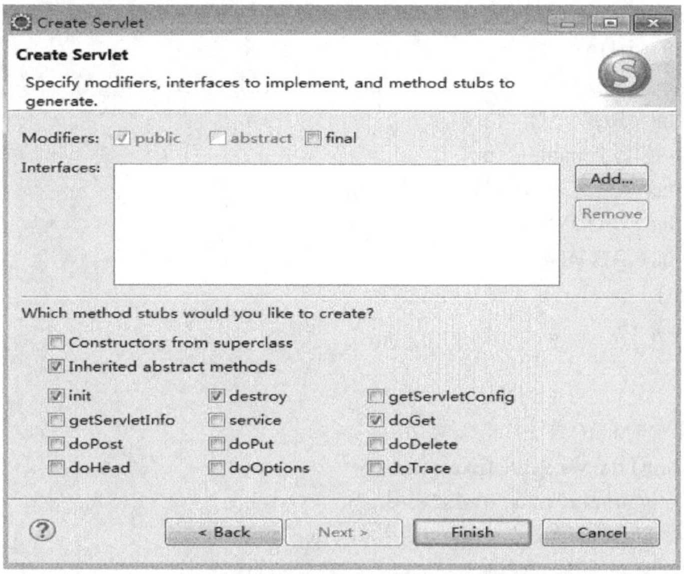

图 4-6

(4) 单击“Finish”按钮完成 Servlet 的创建。编写代码如示例 4-2 所示。

示例 4-2：

```
package control;
```

```
import java.io.IOException;
import java.io.PrintWriter;

import javax.servlet.ServletException;
import javax.servlet.http.HttpServlet;
import javax.servlet.http.HttpServletRequest;
import javax.servlet.http.HttpServletResponse;
import javax.servlet.ServletConfig;

public class ShowCycleServlet extends HttpServlet {

    private ServletConfig config ; // 声明 ServletConfig 引用
    private String name ;
    public void destroy() {
        super.destroy();
        System.out.println("destroy()方法被调用...");
    }

    public void doGet(HttpServletRequest request, HttpServletResponse response)
            throws ServletException, IOException {
    //这里需要修改，否则，中文将是乱码
        response.setContentType("text/html;charset=UTF-8");
        PrintWriter out = response.getWriter();
        System.out.println("客户端 GET 请求一次...");
        out.println("<HTML>");
        out.println("<HEAD><TITLE>第一个 Servlet</TITLE></HEAD>");
        out.println("<BODY>");
        out.print("你好!"+name+"<br>");
        out.println("欢迎进入 Java Web 世界！ ");
        out.println("</BODY>");
        out.println("</HTML>");
        out.flush();
        out.close();
    }

        //重写无参的 init()方法
    public void init() throws ServletException {
         System.out.println("init()方法被调用...");
         config = this.getServletConfig();//获取 ServletConfig 对象
         name = config.getInitParameter("name");//获取初始化参数
      }

// 也可以重写有参数的 init()方法
/*
public void init(ServletConfig config) throws ServletException {
        super.init(config); //必须调用父类的 init()方法
```

```
            System.out.println("init(ServletConfig config)方法被调用...");
            name = config.getInitParameter("name");//获取初始化参数
              }
}
   */
```

(5) 修改 web.xml，加入初始化参数 name，代码如示例 4-3 所示。

示例 4-3：

```
<?xml version="1.0" encoding="UTF-8"?>
<web-app xmlns:xsi="http://www.w3.org/2001/XMLSchema-instance"
xmlns="http://xmlns.jcp.org/xml/ns/javaee"
xsi:schemaLocation="http://xmlns.jcp.org/xml/ns/javaee
http://xmlns.jcp.org/xml/ns/javaee/web-app_3_1.xsd"
id="WebApp_ID" version="3.1">
<display-name>MyFirstWeb</display-name>
<servlet>
      <servlet-name>ShowCycleServlet</servlet-name>
      <servlet-class>control.ShowCycleServlet</servlet-class>
      <!--这里添加初始化参数 -->
      <init-param>
            <param-name>name</param-name>
            <param-value>HOPU</param-value>
      </init-param>
</servlet>
<servlet-mapping>
      <servlet-name>ShowCycleServlet</servlet-name>
      <url-pattern>/showCycleServlet</url-pattern>
</servlet-mapping>
<!-- 这里是欢迎页面 -->
<welcome-file-list>
      <welcome-file>index.jsp</welcome-file>
</welcome-file-list>
</web-app>
```

(6) 部署该 Web 应用到服务器，并启动 Tomcat，在浏览器的地址栏输入“http://localhost:8080/MyFirstWeb/showCycleServlet”，页面显示如图 4-7 所示。此时 Tomcat 服务器的控制台显示如图 4-8 所示。

图 4-7

```
Markers  Properties  Servers  Data Source Explorer  Snippets  Console ⊠  Maven Repositories  Terminal
<terminated> Tomcat v8.5 Server at localhost [Apache Tomcat] D:\Program Files\Java\jdk1.8.0_161\bin\javaw.exe (2018年8月22日 下午4:10:57)
信息: Server startup in 17749 ms
init()方法被调用...
init(ServletConfig config)方法被调用...
客户端GET请求一次...
客户端GET请求一次...
客户端GET请求一次...
八月 22, 2018 4:13:40 下午 org.apache.catalina.core.StandardServer await
信息: A valid shutdown command was received via the shutdown port. Stopping the Server instance.
八月 22, 2018 4:13:40 下午 org.apache.coyote.AbstractProtocol pause
信息: Pausing ProtocolHandler ["http-nio-8080"]
八月 22, 2018 4:13:40 下午 org.apache.coyote.AbstractProtocol pause
信息: Pausing ProtocolHandler ["ajp-nio-8009"]
八月 22, 2018 4:13:40 下午 org.apache.catalina.core.StandardService stopInternal
信息: Stopping service [Catalina]
destroy()方法被调用...
八月 22, 2018 4:13:40 下午 org.apache.catalina.core.ApplicationContext log
信息: SessionListener: contextDestroyed()
八月 22, 2018 4:13:40 下午 org.apache.catalina.core.ApplicationContext log
```

图 4-8

在页面上不停地单击“刷新”按钮，可以看见控制台不停地输出“客户端 GET 请求一次”，但“init()方法被调用”只输出一次，说明第一次请求该 Servlet 时，Servlet 容器创建一个 Servlet 实例，并调用 init()方法来初始化该实例。当 Servlet 实例被创建并装载后，如果有客户端继续请求该 Servlet，容器将不会再创建 Servlet 实例而是使用原来的实例为客户端提供服务，这就是我们常说的 Servlet 实例的单例模式。当停止服务器时，控制台输出“destroy()方法被调用”，说明 Servlet 实例被销毁，占用的资源被释放。

4.5.4 Servlet 应用实例一

在本例中，使用 Eclipse 建立一个简单的 Web 应用，完成的功能是：在一个登录的 JSP 页面中有两个文本框，用户可以输入登录名和密码，可以提交该页面的数据到 Servlet，Servlet 验证登录名和密码。如果用户名和密码匹配，则提示登录成功，否则提示失败。

具体的步骤如下。

(1) 按照 4.5.3 节例子的方式，建立一个新的工程，取名为 LoginWebApp。修改其自动生成的 index.jsp 文件，实现登录页面。代码如示例 4-4 所示。

示例 4-4：

```
<%@ page contentType="text/html; charset=UTF-8"%>
<html>
    <head>
        <title>登录页面</title>
    </head>
    <body>
        <h2>
            欢迎登录
        </h2>
        <form action="mysecondservlet" method="post">
            用户名：
            <input type="text" name="name" value="">
            <br>
            密  码：
```

```
                <input type="password" name="pwd" value="">
                <br>
                <input type="submit" value="提交">
            </form>
        </body>
</html>
```

(2) 建立一个 LoginServlet，处理表单提交的数据，并验证用户名和密码，给出不同的提示。用户名和密码匹配，显示欢迎界面，如果不匹配，显示错误页面。代码如示例 4-5 所示。

示例 4-5：

```
package control;

import java.io.IOException;
import java.io.PrintWriter;
import javax.servlet.ServletException;
import javax.servlet.http.HttpServlet;
import javax.servlet.http.HttpServletRequest;
import javax.servlet.http.HttpServletResponse;
import java.util.Enumeration;

public class LoginServlet extends HttpServlet {

    public void doPost(HttpServletRequest request, HttpServletResponse response)
            throws ServletException, IOException {
        response.setContentType("text/html;charset=UTF-8");
        PrintWriter out = response.getWriter();
        // 获取从表单来的数据
        String name = request.getParameter("name");
        String pwd = request.getParameter("pwd");
        out.println("<HTML>");
        out.println("<HEAD><TITLE>验证结果页面</TITLE></HEAD>");
        out.println("<BODY>");
        // 验证用户名和密码
        if (name.equals("hopu") && pwd.equals("123456")) {
            out.println("欢迎您！ " + name);
        } else {
            out.print("用户名或密码不正确！ ");
        }
        out.println("</BODY>");
        out.println("</HTML>");
        out.flush();
        out.close();
    }
}
```

(3) 发布该 Web 应用，启动 Tomcat 服务器后，打开浏览器，在地址栏输入“http://localhost:8080/LoginWebApp/”，显示如图 4-9 所示的界面。

图 4-9

(4) 分别输入正确的和不存在的用户名和密码，则显示如图 4-10 和图 4-11 所示的界面。

图 4-10

图 4-11

4.5.5 Servlet 应用实例二

在学习了上面的应用实例后，本例进一步采用 JDBC+Servlet+JSP 通过对学生的增删改查进一步巩固 Servlet 知识，以及对三层架构有一定的认识。

具体的步骤如下。

(1) 新建一个 Dynamic Web Project 工程，取名为 Student_CRUD。新建一个学生实体类，放在 domain 包下。代码如示例 4-6 所示。

示例 4-6：

```
package domain;

public class Student {
private Long id;
private String name;
```

```
    private Integer age;
    /*空参构造*/
    public Student() {
        super();
    }
    /*有参构造*/
    public Student(String name, Integer age) {
        super();
        this.name = name;
        this.age = age;
    }
    /*get()、set()方法*/
    public Long getId() {
        return id;
    }

    public void setId(Long id) {
        this.id = id;
    }

    public String getName() {
        return name;
    }

    public void setName(String name) {
        this.name = name;
    }
    public Integer getAge() {
        return age;
    }

    public void setAge(Integer age) {
        this.age = age;
    }
    /*重写 toString()方法*/
    @Override
    public String toString() {
        return "Student [id=" + id + ", name=" + name + ", age=" + age + "]";
    }
}
```

(2) 设计一个 JDBCUtil 工具类，放在 util 包下，方便对数据库的维护处理。代码如示例 4-7 所示。

示例 4-7：

```
package util;
```

```
import java.sql.Connection;
import java.sql.ResultSet;
import java.sql.SQLException;
import java.sql.Statement;
import java.util.Properties;

import javax.sql.DataSource;

import com.alibaba.druid.pool.DruidDataSourceFactory;

public class JDBCUtil {
private static DataSource ds = null;
static {
    // 当 JDBCUtil 执行后，直接加载至 JVM 立即执行
    try {
        Properties p = new Properties();

p.load(Thread.currentThread().getContextClassLoader().getResourceAsStream("jdbc.properties"));
        ds = DruidDataSourceFactory.createDataSource(p);
    } catch (Exception e) {
        e.printStackTrace();
    }
}

public static Connection getConn() {
    try {
        return ds.getConnection();
    } catch (SQLException e) {
        e.printStackTrace();
    }
    return null;
}

public static void close(Connection conn, Statement st, ResultSet rs) {
    try {
        if (rs != null) {
            rs.close();
        }
    } catch (Exception e) {
    } finally {
        try {
            if (st != null) {
                st.close();
            }
        } catch (Exception e) {
            e.printStackTrace();
        } finally {
```

```
                    try {
                        if (conn != null) {
                            conn.close();
                        }
                    } catch (Exception e) {
                        e.printStackTrace();
                    }
                }
            }
        }
    }
```

(3) 数据库的信息传输到页面进行渲染，我们先编写 Dao 层，写接口 IStudentDao，然后完成实现类 IStudentDaoImpl，包含增删改查操作。代码如示例 4-8 和示例 4-9 所示。

示例 4-8：

```
package dao;

import java.util.List;

import domain.Student;

public interface IStudentDao {
/**
 * 保存操作
 *
 * @param stu 学生对象，封装了需要保存的对象
 */
void save(Student stu);

/**
 * 删除操作
 *
 * @param id 被删除学生的主键操作
 */
void delete(Long id);

/**
 *
 * @param id         被更改学生的主键值
 * @param newStu 学生新的信息
 */
void update(Student newStu);

/**
 * 查询指定 id 的学生对象
 *
```

```
 * @param id 被查询学生的主键值
 * @return 如果 id 存在，返回学生对象，否则为 null
 */
Student get(Long id);

/**
 * 查询并返回所有学生对象
 *
 * @return 如果结果集为空，返回一个空的 list 对象
 */
List<Student> listall();
}
```

示例 4-9：

```
package dao;

import java.sql.Connection;
import java.sql.PreparedStatement;
import java.sql.ResultSet;
import java.sql.SQLException;
import java.util.ArrayList;
import java.util.List;

import domain.Student;
import util.JDBCUtil;

public class IStudentDaoImpl implements IStudentDao {

public void save(Student stu) {
	String sql = "insert into t_student (name,age) values (?,?)";
	PreparedStatement ps = null;
	Connection conn = null;
	try {
		conn = JDBCUtil.getConn();
		ps = conn.prepareStatement(sql);// 预处理
		ps.setString(1, stu.getName());
		ps.setInt(2, stu.getAge());
		ps.executeUpdate();// 执行
	} catch (SQLException e) {
		e.printStackTrace();
	}
	JDBCUtil.close(conn, ps, null);// 事务

}

@Override
```

```
public void delete(Long id) {
    String sql = "delete from t_student where id = ?";
    PreparedStatement ps = null;
    Connection conn = null;
    try {
        conn = JDBCUtil.getConn();
        ps = conn.prepareStatement(sql);
        ps.setLong(1, id);
        ps.executeUpdate();
    } catch (Exception e) {
        e.printStackTrace();
    }
    JDBCUtil.close(conn, ps, null);
}

@Override
public void update(Student stu) {
    String sql = "update t_student set name =? ,age=? where id=?";
    PreparedStatement ps = null;
    Connection conn = null;
    try {

        conn = JDBCUtil.getConn();
        ps = conn.prepareStatement(sql);
        ps.setString(1, stu.getName());
        ps.setInt(2, stu.getAge());
        ps.setLong(3, stu.getId());
        ps.executeUpdate();
    } catch (Exception e) {
        e.printStackTrace();
    }
    JDBCUtil.close(conn, ps, null);
}

public Student get(Long id) {
    String sql = "select * from t_student where id=?";
    PreparedStatement ps = null;
    Connection conn = null;
    ResultSet rs = null;
    try {
        conn = JDBCUtil.getConn();
        ps = conn.prepareStatement(sql);
        ps.setLong(1, id);
        rs = ps.executeQuery();
        if (rs.next()) {
            Student stu = new Student();
            stu.setId(rs.getLong("id"));
```

```
                stu.setName(rs.getString("name"));
                stu.setAge(rs.getInt("age"));
                return stu;
            }
        } catch (Exception e) {
            e.printStackTrace();
        }
        JDBCUtil.close(conn, ps, rs);
        return null;
    }

    @Override
    public List<Student> listall() {
        List<Student> list = new ArrayList<>();
        String sql = "select * from t_student";
        PreparedStatement ps = null;
        Connection conn = null;
        ResultSet rs = null;
        try {
            conn = JDBCUtil.getConn();
            ps = conn.prepareStatement(sql);
            rs = ps.executeQuery();
            while (rs.next()) {
                Student stu = new Student();
                stu.setId(rs.getLong("id"));
                stu.setName(rs.getString("name"));
                stu.setAge(rs.getInt("age"));
                list.add(stu);
            }
        } catch (Exception e) {
            e.printStackTrace();
        } finally {
            JDBCUtil.close(conn, ps, rs);
        }
        return list;

    }
}
```

(4) 编写 ListStudentServlet 类并放在 control 包下，用于展示所有用户信息。代码如示例 4-10 所示。虽然我们目前还没有系统性地学习过注解，但在这里可以简单地初步认识了解一下，在 Servlet 3.0 以后，我们可以不在 web.xml 中配置 servlet，只需要加上@WebServlet 注解就可以修改该 servlet 的属性了。@WebServlet("/xx/yy")的作用就是当浏览器访问带有以/xx/yy 结尾的路径时，若没有指定 Servlet 的名称，默认以类名作为 Servlet 对应的名字。相对应的 Servlet 就会进行拦截。例如在示例 4-10 中，你的访问路径为 http://localhost:8080/Student_CRUD/student/list 时，就会被 ListStudentServlet 拦截。采用注解后，可以使我们不用再去在 web.xml 中进行一系列较烦琐的配置了。

示例 4-10：

```
package control;

import java.io.IOException;
import java.util.List;

import javax.servlet.ServletException;
import javax.servlet.annotation.WebServlet;
import javax.servlet.http.HttpServlet;
import javax.servlet.http.HttpServletRequest;
import javax.servlet.http.HttpServletResponse;

import dao.IStudentDao;
import dao.IStudentDaoImpl;
import domain.Student;

@WebServlet("/student/list")
public class ListStudentServlet extends HttpServlet {
private static final long serialVersionUID = 1L;
private IStudentDao dao;

@Override
public void init() throws ServletException {
	dao = new IStudentDaoImpl();
}

@Override
protected void service(HttpServletRequest req, HttpServletResponse resp) throws ServletException,
IOException {
	// 1.接收请求参数封装成对象

	// 2.调用业务方法处理请求
	List<Student> list = dao.listall();
	req.setAttribute("students", list);

	// 3.控制界面跳转
	req.getRequestDispatcher("/WEB-INF/view/list.jsp").forward(req, resp);

 }
}
```

(5) 在 WEB-INF 下创建 list.jsp 页面。通过${student}标签获取后端传输过来的数据。

然后使用 c 标签的 forEach 遍历，动态展示学生数据，关于 JSP 的标签库内容将在后面的章节中进行学习。代码如示例 4-11 所示。

示例 4-11：

```
<%@ page language="java" contentType="text/html; charset=UTF-8"
    pageEncoding="UTF-8"%>
    <%@ taglib prefix="c" uri="http://java.sun.com/jsp/jstl/core" %>
<!DOCTYPE html>
<html>
<head>
<meta charset="UTF-8">
<title>学生信息</title>
</head>
<body>
<h2 align="center"> 学生列表</h2>
    <a href="/Student_CRUD/student/edit">添加学生</a>
     <table border="1" width="50%" cellpadding="0" cellspacing="0" align="center">
            <tr style="background-color: orange;">
                <th>编号</th>
                <th>姓名</th>
                <th>年龄</th>
                <th>操作</th>
            </tr>

     <c:forEach items="${students}" var="s" varStatus="vs">
            <tr style='background-color: ${vs.count %2==0 ?"LavenderBlush":"gray"};'>
                <td>${s.id }</td>
                <td>${s.name }</td>
                <td>${s.age }</td>
                <td>
                    <a href="/Student_CRUD/student/delete?id=${s.id}">删除</a>       |
                    <a href="/Student_CRUD/student/edit?id=${s.id}">编辑</a>
                </td>
            </tr>
    </c:forEach>
     </table>
</body>
</html>
```

在浏览器栏输入 http://localhost:8080/Student_CRUD/student/list，运行结果如图 4-12 所示。

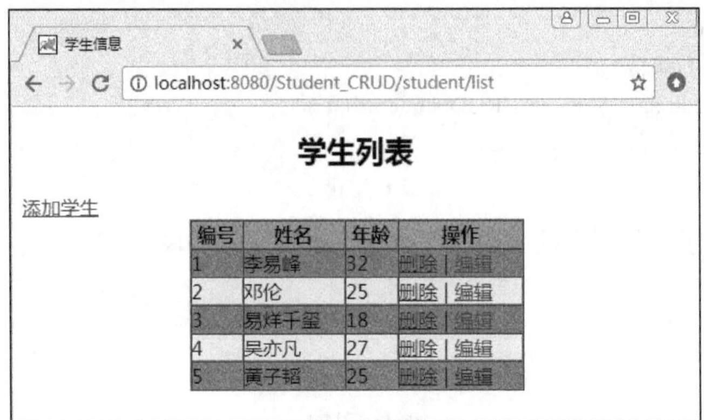

图 4-12

(6) 同理，在 control 包下新建一个 DeleteStudentServlet 类，用来处理删除操作。删除完成后，再次调转到展示页面。代码如示例 4-12 所示。

示例 4-12：

```
package control;

import java.io.IOException;

import javax.servlet.ServletException;
import javax.servlet.annotation.WebServlet;
import javax.servlet.http.HttpServlet;
import javax.servlet.http.HttpServletRequest;
import javax.servlet.http.HttpServletResponse;

import dao.IStudentDao;
import dao.IStudentDaoImpl;

@WebServlet("/student/delete")
public class DeleteStudentServlet extends HttpServlet {
private static final long serialVersionUID = 1L;
private IStudentDao dao;

@Override
public void init() throws ServletException {
        dao = new IStudentDaoImpl();
}
@Override
protected void service(HttpServletRequest req, HttpServletResponse resp) throws ServletException,
IOException {
        // 1.接收请求参数封装成对象
        Long id = Long.valueOf(req.getParameter("id"));
        // 2.调用业务方法处理请求
```

```
        dao.delete(id);
        // 3.控制界面跳转
        resp.sendRedirect("/Student_CRUD/student/list");
    }
}
```

单击 5 号“黄子韬”栏中的“删除”按钮，运行结果如图 4-13 所示，5 号已经被删除。

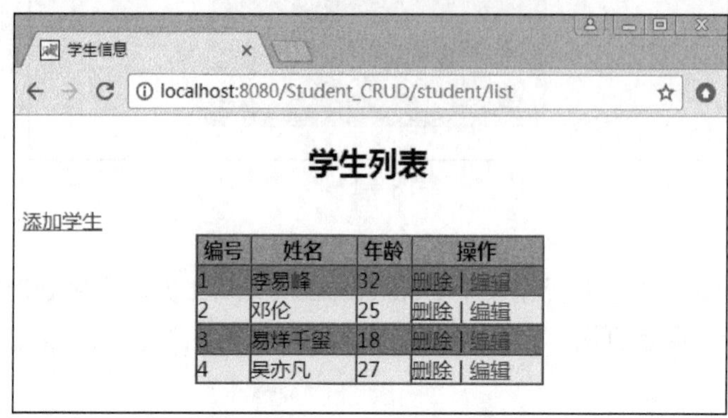

图 4-13

(7) 在 control 包下新建一个 EditStudentServlet 类，用来处理修改和添加操作，修改操作需要回显学生信息，因此需要判断数据是否有 id 值，当没有 id 值时就是添加操作，有 id 值时就是修改操作，此时需要获取学生信息，然后传递给页面进行回显。代码如示例 4-13 所示。

示例 4-13：

```
package control;

import java.io.IOException;

import javax.servlet.ServletException;
import javax.servlet.annotation.WebServlet;
import javax.servlet.http.HttpServlet;
import javax.servlet.http.HttpServletRequest;
import javax.servlet.http.HttpServletResponse;

import dao.IStudentDao;
import dao.IStudentDaoImpl;
import domain.Student;

@WebServlet("/student/edit")
public class EditStudentServlet extends HttpServlet {
    private static final long serialVersionUID = 1L;
    private IStudentDao dao;

    @Override
```

```
    public void init() throws ServletException {
        dao = new IStudentDaoImpl();
    }

    protected void service(HttpServletRequest req, HttpServletResponse resp) throws ServletException,
    IOException {
        // 1.接收请求参数封装成对象
        String sid = req.getParameter("id");

        // 2.调用业务方法处理请求
        if (hasLength(sid)) {
            Student stu = dao.get(Long.valueOf(sid));
            req.setAttribute("student", stu);// 传递给 edit.jsp，用于回显被编辑的学生
        }
        // 3.控制界面跳转
        req.getRequestDispatcher("/WEB-INF/view/edit.jsp").forward(req, resp);
    }

    private boolean hasLength(String str) {
        return str != null && !"".equals(str.trim());
    }
}
```

(8) 在 WEB-INF 下创建 edit.jsp 页面，用于编辑学生信息。代码如示例 4-14 所示。

示例 4-14：

```
<%@ page language="java" contentType="text/html; charset=UTF-8"
    pageEncoding="UTF-8"%>
<!DOCTYPE html>
<html>
<head>
<meta charset="UTF-8">
<title>修改页面</title>
</head>
<body>
<form action="/Student_CRUD/student/save" method="post">
        <input type="hidden" name="id" value="${student.id }">
        姓名<input type="text" name="name" required value="${student.name }"/><br/>
        年龄<input type="number" name="age" required value="${student.age }"><br/>

        <input type="submit" value="${student==null ?"保存学生信息":"修改学生信息"}">
    </form>
</body>
</html>
```

(9) 新建一个 SaveStudentServlet 类，同理也是根据数据是否有 id 值来判断是更新操作，还是新增学生信息。代码如示例 4-15 所示。

示例 4-15：

```
package control;

import java.io.IOException;

import javax.servlet.ServletException;
import javax.servlet.annotation.WebServlet;
import javax.servlet.http.HttpServlet;
import javax.servlet.http.HttpServletRequest;
import javax.servlet.http.HttpServletResponse;

import dao.IStudentDao;
import dao.IStudentDaoImpl;
import domain.Student;

@WebServlet("/student/save")
//保存学生信息
public class SaveStudentServlet extends HttpServlet {
    private static final long serialVersionUID = 1L;
    private IStudentDao dao;

    @Override
    public void init() throws ServletException {
        dao = new IStudentDaoImpl();
    }

    protected void service(HttpServletRequest req, HttpServletResponse resp) throws ServletException,
    IOException {
        // 1.接收请求参数封装成对象
        req.setCharacterEncoding("UTF-8");
        String name = req.getParameter("name");
        Integer age = Integer.valueOf(req.getParameter("age"));
        Student stu = new Student(name, Integer.valueOf(age));
        // 2.调用业务方法处理请求
        String id = req.getParameter("id");
        if (hasLength(id)) {// 更新保存
            stu.setId(Long.valueOf(id));
            dao.update(stu);
        } else { // 新增
            dao.save(stu);
        }
        // 3.控制界面跳转
        resp.sendRedirect("/Student_CRUD/student/list");

    }

    private boolean hasLength(String str) {
```

```
        return str != null && !"".equals(str.trim());
    }
}
```

单击 1 号“李易峰”栏中的“编辑”按钮，页面跳转到编辑页，修改年龄为 18，页面如图 4-14 所示。单击“修改学生信息”按钮，页面跳转到展示页面，最终页面如图 4-15 所示。

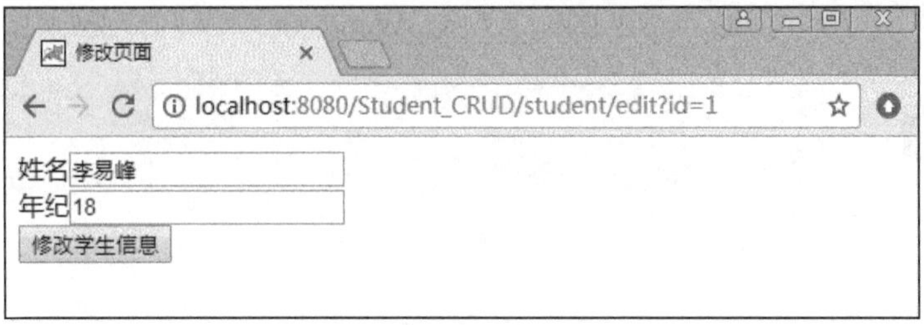

修改页面
localhost:8080/Student_CRUD/student/edit?id=1
姓名 李易峰
年纪 18
修改学生信息

图 4-14

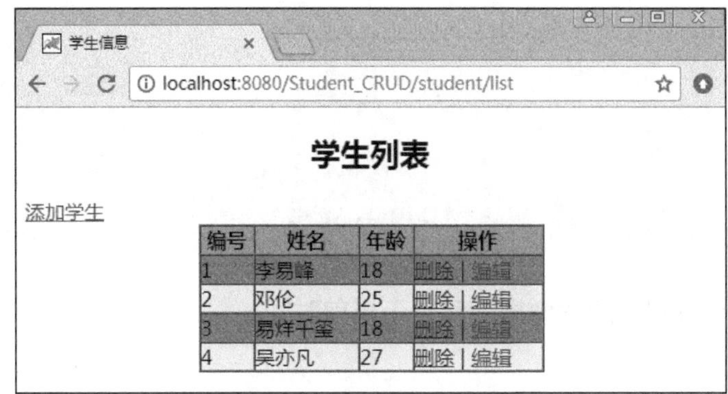

图 4-15

同样单击“添加学生”按钮，页面跳转到编辑页，输入学生信息。然后单击“修改学生信息”按钮，最终跳转到展示页面重新展示所有学生信息，操作步骤与修改类似，在此不再赘述。

【单元小结】

- Servlet 是在服务器上运行的 Java 代码。
- 自定义的 Servlet 扩展自 HttpServlet 类，一般需要重写 doGet()和 doPost()方法。
- Servlet 生命周期经历 3 个阶段，init()、service()、destroy()方法会被调用。
- 客户提交的表单参数可使用 request.getParameter("")方法来获取。
- 使用 Eclipse 开发完 Web 应用必须发布到 Web 服务器中。

【单元自测】

1. 在表单中，用于发送敏感数据的方法是(　　)。

A. POST　　B. GET　　C. PUT　　D. OPTIONS

2. Web 应用程序结构中，web.xml 文件必须放在以下哪个目录下？(　　)

A. 工程　　B. WEB-INF　　C. classes　　D. src

3. 在 Eclipse 开发环境中，网页文件放在以下哪个目录中？(　　)

A. 工程　　B. WEB-INF　　C. WebRoot　　D. classes

4. 在 Eclipse 开发环境中，.java 文件放在以下哪个目录中？(　　)

A. 工程　　B. WEB-INF　　C. WebRoot　　D. Src

【上机实战】

上机目标

- 掌握 Servlet API 中的方法。
- 掌握在 Eclipse 中开发 Servlet 的方法。

上机练习

◆ 第一阶段 ◆

练习 1：演示 Servlet 的生命周期和显示当前时间

【问题描述】

使用 Servlet API 来开发一个简单的应用程序，显示当前的时间。同时，当用户单击浏览器的“刷新”按钮时，当前的时间会更新。

【问题分析】

(1) 需要建立一个 ShowTime 类，该类的实例对象将生成时间的字符串。

(2) 编写一个 Servlet，其中的 doPost()方法负责显示时间信息。

【参考步骤】

(1) 新建一个 Web 工程，取名为 ServletDemo1。

(2) 在 Eclipse 中新建一个 ShowTime 类，放在 myutil 包下面。在编辑器中编写 ShowTime 类的代码，如示例 4-16 所示。

示例 4-16：

```
package myutil;
import java.text.SimpleDateFormat;
import java.util.Date;

public class ShowTime {
   private Date date;
```

```
    private SimpleDateFormat dateFormat;
    public ShowTime() {
        date = new Date();
        dateFormat = new SimpleDateFormat("现在时间是:yyyy-MM-dd hh:mm:ss");
    }
    public String getTime() {
        return dateFormat.format(date);
    }
}
```

(3) 新建一个名为 ShowTimeServlet 的 Servlet。代码如示例 4-17 所示。

示例 4-17：

```
package control;

import java.io.IOException;
import java.io.PrintWriter;
import myutil.ShowTime;

import javax.servlet.ServletConfig;
import javax.servlet.ServletException;
import javax.servlet.http.HttpServlet;
import javax.servlet.http.HttpServletRequest;
import javax.servlet.http.HttpServletResponse;

public class ShowTimeServlet extends HttpServlet {
    ShowTime show =null;

    public void init(ServletConfig cnf) throws ServletException {
        super.init(cnf);
        System.out.println("init()方法被调用");
    }

    public void doGet(HttpServletRequest req, HttpServletResponse res)
            throws IOException, ServletException {
        res.setContentType("text/html; charset=gb2312");
        PrintWriter out = res.getWriter();
        show = new ShowTime();
        out.println("<HTML>");
        out.println("<HEAD><TITLE>");
        out.println("显示时间信息");
        out.println("</TITLE></HEAD>");
        out.println("<BODY>");
        out.println("<CENTER>");
        out.println("现在时间是:"+show.getTime());
        out.println("</CENTER>");
        out.println("</BODY></HTML>");
```

```
    }

    public void destroy() {
        System.out.println("destroy()方法被调用了");
    }
}
```

(4) 把该 Web 应用发布到 Tomcat 服务器并启动，打开浏览器，在地址栏中输入"http://localhost:8080/ServletDemo1/showTimeServlet"，将显示如图 4-16 所示的界面。

图 4-16

(5) 查看控制台的输出。

◆ 第二阶段 ◆

练习 2：改进前面的 LoginWebApp 工程，实现能真正和数据库交互的用户验证

【问题描述】

编写一个 JSP 页面，接受用户输入的用户名和密码，使用 Servlet 来处理用户的请求，如果数据库表 UserInfo 中存在该用户，则显示欢迎界面，否则显示用户名或密码错误界面。

【问题分析】

(1) 需要建立一个 JSP 页面，用户可以输入用户名和密码。

(2) 使用三层结构的形式来完成数据库的操作。

(3) 编写一个 Servlet，其中的 doPost()方法负责处理 JSP 页面的请求，并和数据库交互。

【参考步骤】

(1) 新建一个 Web 工程，取名为 ServletDemo2。

(2) 改写 index.jsp 页面，代码如示例 4-18 所示。

示例 4-18：

```
<%@ page contentType="text/html; charset=UTF-8 "%>
<html>
    <head>
        <title>登录页面</title>
    </head>
    <body>
```

```
        <h3>欢迎登录</h3>
        <form action="logonServlet" method="post">
            请输入你的登录名:
            <input type="text" name="username" value="">
            <br>
            请输入你的密码:
            <input type="password" name="userpwd" value="">
            <br>
            <input type="submit" value="提交">
        </form>
    </body>
</html>
```

(3) 新建 Java 类，取名 DBHelper.java，实现数据库的连接和关闭功能，代码如示例 4-19 所示。

示例 4-19：

```
package dbutils;

import java.io.IOException;
import java.sql.Connection;
import java.sql.DriverManager;
import java.util.Properties;
import java.io.InputStream;

/**
 * 提供连接，关闭连接，针对 service
 * @author svse
 */
public class DBHelper {
    // 属性字符串
    private static String driver;
    private static String url;
    private static String username;
    private static String password;
    private static Connection conn;

    /**
     * 初始化各种 JDBC 数据库连接参数
     */
    private static void init(){
        try {
            // 创建属性集对象
    Properties props = new Properties();
    InputStream fis =DBHelper.class.getResourceAsStream("/jdbc.properties");
props.load(fis);
    driver = props.getProperty("driver");
```

```
            url = props.getProperty("url");
            username = props.getProperty("username");
            password = props.getProperty("password");
        } catch (IOException e) {
            e.printStackTrace();
        }
    }
    /**
     * 提供连接
     * @return 连接
     */
    public static Connection getConn() {
        init();
        try {
            Class.forName(driver);
            conn = DriverManager.getConnection(url, username, password);
        } catch (Exception e) {
            e.printStackTrace();
        }
        return conn;
    }
    /**
     * 关闭连接
     */
    public static void closeConn(){
        try {
            if(conn!=null){
                conn.close();
                conn = null;
            }
        } catch (Exception e) {
            e.printStackTrace();
        }
    }
}
```

(4) 这里用到了属性文件 jdbc.properties。新建一个属性文件，放在 src 的目录下，文件中使用键值对的形式来存放数据库驱动、URL、数据库用户名和密码。代码如示例 4-20 所示。

示例 4-20：

```
driver=com.mysql.jdbc.Driver
url=jdbc:mysql://localhost:3306/myweb?useUnicode=true&characterEncoding=utf-8
username=root
password=root
```

(5) 在 support 包下再建立 DBCommand.java 类，代码如示例 4-21 所示。

示例 4-21：

```
package support;

import java.sql.PreparedStatement;
import java.sql.ResultSet;
import java.sql.ResultSetMetaData;
import java.util.ArrayList;
import java.util.HashMap;
import java.util.Iterator;
import java.util.List;
import java.util.Map;

/**
 * 所有操作的入口，针对 DAO，提供了关闭 pstm、rs 的方法
* @author svse
*/
public class DBCommand {
    private static ResultSet rs;

    private DBCommand() {
    }
    /**
     * 增删改的操作入口
     * @param pstm
     * @param paramsMap
     * @return 操作是否成功对应的数字，-1 表示出现异常
     */
    public static int execUpdate(PreparedStatement pstm,
        Map<Object, Object> paramsMap) {
        try {
            Iterator<Object> it = paramsMap.keySet().iterator();
            int i = 1;
            while (it.hasNext()) {
                pstm.setObject(i++, paramsMap.get(it.next()));
            }
            return pstm.executeUpdate();
        } catch (Exception e) {
            e.printStackTrace();
        }finally{
            close(pstm);
        }
        return - 1;
    }
```

```
/**
 * 带参数的查询入口
 * 返回值是 List，内含多个 map，一个 map 代表一行
 * @param pstm
 * @param paramsMap
 * @return List
 */
public static List<Map<String, Object>> execQuery(PreparedStatement pstm,
    Map<Object, Object> paramsMap) {
    try {
        Iterator<Object> it = paramsMap.keySet().iterator();
        int i = 1;
        while (it.hasNext()) {
            pstm.setObject(i++, paramsMap.get(it.next()));
        }
        rs = pstm.executeQuery();
        return readRs(rs);
    } catch (Exception e) {
        e.printStackTrace();
    }finally{
        close(rs);
        close(pstm);
    }
    return null;
}

/**
 * 无参数的查询入口
 * 返回值是 List，内含多个 map，一个 map 代表一行
 * @param pstm
 * @return list
 */
public static List<Map<String, Object>> execQuery(PreparedStatement pstm) {
    try {
        rs = pstm.executeQuery();
        return readRs(rs);
    } catch (Exception e) {
        e.printStackTrace();
    }finally{
        close(rs);
        close(pstm);
    }
    return null;
}
/**
 * 提供对 ResultSet 的解析
 * 将结果集的每行解析为 map，放入 list
```

```
     * @param rs
     * @return list
     * @throws Exception
     */
    public static List<Map<String, Object>> readRs(ResultSet rs)
        throws Exception {
        List<Map<String, Object>> data = new ArrayList<Map<String, Object>>();
        //获得列名组合
        ResultSetMetaData rsmd = rs.getMetaData();
        //获得列的个数
        int columnCount = rsmd.getColumnCount();
        while (rs.next()) {
            Map<String, Object> row = new HashMap<String, Object>();
            for (int index = 1; index <= columnCount; index++) {
                String columnName = rsmd.getColumnName(index);
                row.put(columnName, rs.getObject(columnName));
            }
            data.add(row);
        }
        return data;
    }

    /**
     * 关闭语句
     * @param pstm
     */
    public static void close(PreparedStatement pstm) {
        try {
            if (pstm != null) {
                pstm.close();
                pstm = null;
            }
        } catch (Exception ex) {
            ex.printStackTrace();
        }
    }

    /**
     * 关闭结果集
     * @param rs
     */
    public static void close(ResultSet rs) {
        try {
            if (rs != null) {
                rs.close();
                rs = null;
            }
```

```
            } catch (Exception ex) {
                ex.printStackTrace();
            }
        }
    }
```

(6) 上面的类是数据库操作的基础类，接着新建与数据库表对应的实体类 UserInfoBean.java，代码如示例 4-22 所示。

示例 4-22：

```
public class UserInfoBean {

    private Integer id;
    private String name;
    private String password;

    public UserInfoBean() {
    }

    public Integer getId() {
        return id;
    }

    public String getName() {
        return name;
    }

    public void setName(String name) {
        this.name = name;
    }

    public String getPassword() {
        return password;
    }

    public void setPassword(String password) {
        this.password = password;
    }

    public void setId(Integer id) {
        this.id = id;
    }
}
```

(7) 建立 BaseDAO.java、UserInfoDAO.java 两个抽象类和 UserInfoDAOImpl.java 实现类来完成对数据库表 UserInfo 的基本操作。代码分别如示例 4-23、示例 4-24 和示例 4-25 所示。

示例 4-23：

```
package dao;

import java.sql.Connection;

public abstract class BaseDAO {
    private Connection conn;
    public void setConn(Connection conn){
        this.conn = conn;
    }
    public Connection getConn(){
        return conn;
    }
}
```

示例 4-24：

```
package dao;

import bean.UserInfoBean;

public abstract class UserInfoDAO extends BaseDAO {

    public abstract UserInfoBean getUserInfoByNameAndPwd(String name,
String password);
}
```

示例 4-25：

```
package dao;

import java.sql.PreparedStatement;
import java.util.LinkedHashMap;
import java.util.List;
import java.util.Map;
import support.DBCommand;
import bean.UserInfoBean;

public class UserInfoDAOImpl extends UserInfoDAO {

    private PreparedStatement pstm;

    public UserInfoBean getUserInfoByNameAndPwd(String name, String password) {
        UserInfoBean user = null;
        try {
pstm = this.getConn().prepareStatement("select * from userinfo where
```

```
username = ? and userpwd = ?");
                    Map<Object,Object> paramsMap = new LinkedHashMap<Object,Object>();
                    paramsMap.put("userName", name);
                    paramsMap.put("userPwd", password);
List<Map<String,Object>> stuList = DBCommand.execQuery(pstm,paramsMap);
            if(stuList.size() != 0){
                    user = new UserInfoBean();
                    Map<String, Object> row = stuList.get(0);
                    System.out.println(row.get("id"));
                    user.setId(new Integer((int) row.get("id")));
                    user.setName(row.get("username").toString());
                    user.setPassword(row.get("userpwd").toString());
            }
        } catch (Exception e) {
            e.printStackTrace();
        }
        return user;
    }
}
```

(8) 建立 UserInfoService.java 接口和 UserInfoServiceImpl.java 实现类来完成业务操作。代码分别如示例 4-26、示例 4-27 所示。

示例 4-26：

```
package service;

import bean.UserInfoBean;

public interface UserInfoService {
    public UserInfoBean getUserInfoByNameAndPwd(String name, String password);
}
```

示例 4-27：

```
package service;

import dao.UserInfoDAO;
import dao.UserInfoDAOImpl;
import support.DBHelper;
import bean.UserInfoBean;

public class UserInfoServiceImpl implements UserInfoService {

    private UserInfoDAO dao = new UserInfoDAOImpl();

    public UserInfoBean getUserInfoByNameAndPwd(String name, String password) {
```

```
            dao.setConn(DBHelper.getConn());
                UserInfoBean userinfo = null;
            try {
                userinfo = dao.getUserInfoByNameAndPwd(name, password);
            } catch (RuntimeException e) {
                e.printStackTrace();
            }finally{
                DBHelper.closeConn();
            }
            return userinfo;
        }
    }
```

(9) 新建一个名为 LoginServlet 的 Servlet，代码如示例 4-28 所示。

示例 4-28：

```
package control;

import java.io.IOException;
import java.io.PrintWriter;
import javax.servlet.ServletException;
import javax.servlet.http.HttpServlet;
import javax.servlet.http.HttpServletRequest;
import javax.servlet.http.HttpServletResponse;

import bean.UserInfoBean;
import dao.UserInfoDAO;
import dao.UserInfoDAOImpl;
import service.UserInfoService;
import service.UserInfoServiceImpl;

public class LogonServlet extends HttpServlet {

    public void doPost(HttpServletRequest request, HttpServletResponse response)
        throws ServletException, IOException {
        // 解决中文问题
        request.setCharacterEncoding("UTF-8");
        response.setContentType("text/html;charset=UTF-8");
        PrintWriter out = response.getWriter();
        // 获取用户提交的参数列表
        String username = request.getParameter("username");
        String userpwd = request.getParameter("userpwd");
        // 生成业务对象
        UserInfoService service = new UserInfoServiceImpl();
        //调用业务方法
    UserInfoBean ubean = service.getUserInfoByNameAndPwd(username,userpwd);
        out.println("<HTML>");
```

```
            out.println("    <HEAD><TITLE>登录 Servlet</TITLE></HEAD>");
            out.println("    <BODY>");
            if (ubean != null) {
                out.print("欢迎你" + ubean.getName());
            } else {
                out.print("用户名或密码不正确");
            }
            out.println("    </BODY>");
            out.println("</HTML>");
            out.flush();
            out.close();
        }
    }
```

(10) 发布该Web应用到Tomcat服务器并启动，打开浏览器，在地址栏输入“http://localhost:8080/ServletDemo2/”，显示如图4-17所示的界面。

图 4-17

(11) 分别输入正确的和不存在的用户名和密码，则分别显示如图 4-18 和图 4-19 所示的界面。

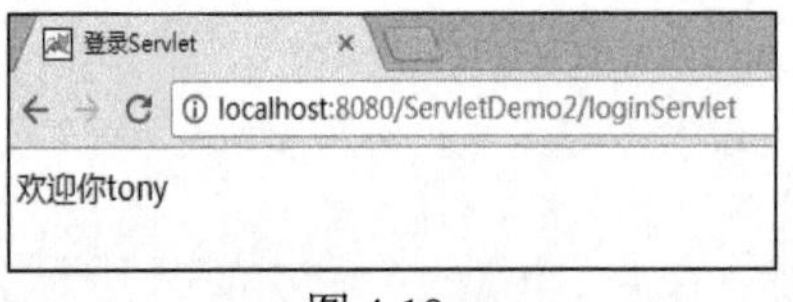

图 4-18

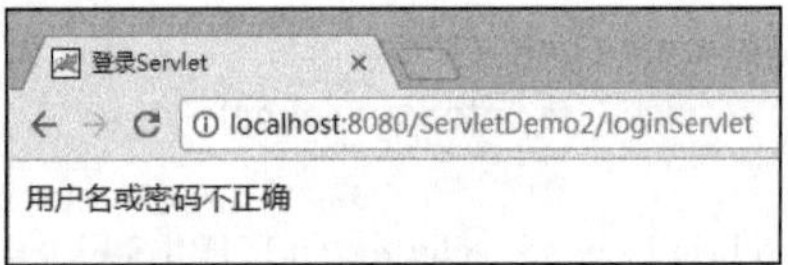

图 4-19

【拓展作业】

1. 编写一个 Servlet，在 doGet()方法中显示一个 Form 表单，用户可以输入姓名和电子邮件地址，用户使用 post 方式提交该表单后，该 Servlet 用 doPost()方法动态读出请求参数，并输出这些参数。

2. 编写一个 Servlet，用来显示远程主机、远程端口、请求的 URI 服务器名称和服务器端口(使用 HttpServletRequest 接口中定义的方法)。

3. 编写一个 Servlet，输出客户端请求的各种标题信息(使用 request 对象的 getHeaderNames()方法)。

单元五

会话管理和使用

课程目标

- ▶ 掌握会话的概念
- ▶ 掌握会话的管理
- ▶ 掌握常用会话跟踪技术
- ▶ 掌握查询字符串的使用

简介

会话跟踪是一种灵活、轻便的机制，它使在页面上的状态编程变为可能。HTTP 是一种无状态协议，每当用户发出请求时，服务器就做出响应，客户端与服务器之间的联系是离散的、非连续的。当用户在同一网站的多个页面之间转换时，根本无法知道是否是同一个客户，会话跟踪就可以解决这个问题。当一个客户在多个页面间切换时，服务器会保存该用户的信息。

5.1 HTTP 协议的无状态

HTTP 是无状态协议。如果一个协议不能记忆它之前的连接，就不能把某客户端的请求与另一个客户端区分开来，我们就说这个协议是无状态的。我们都使用过 FTP(文件传输协议)，它就是一个有状态的协议，原因是连接不会因每次请求而建立和断开。HTTP 协议无状态的性质，使它没有办法跟踪客户端对一个 Web 站点的遍历，客户端的每次请求都使用由匿名用户创建的新连接，这对于需要用户认证的安全站点或为客户提供虚拟购物车的商业站点来说，是很大的挑战，因为状态信息对他们来说是非常有用的。

5.1.1 什么是会话

理解会话前，先来看看 HTTP 处理过程，典型的无状态的处理就像这样：客户端先建立同服务器的一个连接，然后向服务器发送请求，服务器处理请求，然后把响应结果传回给客户端，之后便断开与客户端的连接。如果同一个客户端又再次发送了新请求，服务器并不能把这次的新的连接同此客户端的前一次连接联系起来。因为 HTTP 协议在两次请求的时间间隔里不能保持客户端和服务器之间的持续连接，所以一旦连接断开，服务器和客户端的联系也就结束了。这就是 HTTP 协议面临的巨大挑战，同样也有下面的问题产生。

- 如果服务器需要对客户进行身份验证(即客户必须以有效的用户名和密码登录)，客户端必须对每次请求重新进行身份验证。因为服务器意识不到这个客户已经授权了。
- 由于服务器不能将一个客户和另一个客户区分开来，其结果是不可能存储特定的用户信息，如用户购物车中的内容或者是该用户的喜好。

这些问题可以通过在客户端和服务器端建立一个持续的“虚拟”连接得到解决。虚拟连接把服务器收到的每个请求同发送该请求的客户联系起来。对这种联系的实现可以让客户端返回一些特定的信息给服务器，服务器利用这些特定信息来标识一个客户，并把当前的请求同这个客户以前的请求联系起来。我们把这个虚拟的连接叫作 HTTP 会话，简称会话。

既然会话在请求之间持续存在，可能有人会问“会话什么时候终止”，怎样算是终止了会话?

5.1.2 状态和会话管理

既然说 HTTP 协议是无状态的协议，那么什么是“状态”？我们可以把客户端购物车中的东西或是客户的喜好都叫作状态。同时通过用客户端和服务器端传递特定的客户信息来维持客户的这些状态。这个特定的信息又是什么呢？我们称之为会话 ID，通过这个 ID 来唯一地标识一个客户。

回归到前面的问题，这个会话持续的时间是怎样管理的？我们可以在服务器上进行配置。常见的方法是给会话设定一个时间，叫有效期，超过这个时间，服务器会自动清除这个会话。清除这个会话就意味着把用来识别客户的 ID 连同相关的数据(客户的状态)全部删掉，这就是我们说的会话管理。

5.2 Servlet API 对会话的支持

在 Servlet API 中使用 HttpSession 接口来实现会话，客户端向服务器发送请求，服务器就会创建一个与当前请求相关联的会话对象，并使用 HttpSession 对象封装 HTTP 会话的重要信息，如唯一的会话 ID 以及其他一些特定的客户信息。其中，特定的客户信息可以包含任何 Java 对象。

5.2.1 HttpSession 接口

HttpSession 接口实现了会话机制，并用 Session 来跟踪客户的状态。Session 是在一段时间内，单个客户与 Web 服务器的一连串的相关的交互过程。在一个 Session 中，客户可能会多次请求访问同一个页面，也可能请求访问不同的服务器资源。例如，在网上购物，从客户开始购物直到最后结账，整个过程是一个 Session。Servlet 容器必须要实现这个接口，当一个 Session 开始时，Servlet 容器将创建一个 HttpSession 对象，在该对象中可以保存用户的状态信息。同时，Servlet 容器为 HttpSession 对象分配一个唯一标识符，我们称为 Session ID。Servlet 容器把 Session ID 作为 Cookie 保存在客户的浏览器中。每次客户发出 HTTP 请求时，Servlet 容器可以从 HttpRequest 对象中读取 Session ID，然后根据 Session ID 找到相对应的 HttpSession 对象。

5.2.2 会话对象的创建

在 HttpServletRequest 接口中定义了两种方式来创建会话，分别是 getSession() 和 getSession(boolean create)的方法。这两个方法都能返回与当前请求相关联的HttpSession对象。无参的getSession()方法先查看有没有与当前请求相关联的会话对象，如果有，直接返回该会话对象；如果没有，该方法将会新创建一个会话并返回。而有参的getSession(boolean create)方法将根据布尔值create来决定是否创建会话，如果入参的布尔值是true并且没有与请求相关联的会话，则该方法将会创建一个新的会话并返回；如果有，则直接返回已有的会话对象。如果入参的布尔值是false，并且没有与请求相关联的会话，则该方法不会创建新的会话，直接返回null。

必须确保会话被正确地维护，在程序中，在响应被提交之前必须调用此方法。如果容器需要 Cookie 来维护会话的完整性而会话没有被创建，则会抛出 IllegalStateException 异常。

5.2.3 会话管理

在 Web 应用中，会话必须被很好地管理，HttpSession 接口中提供了管理会话的方法，如表 5-1 所示。

表 5-1

方 法	说 明
String getId()	返回 Session 的 ID
long getCreationTime()	返回会话被创建的时间
long getLastAccessedTime()	返回会话最后处理的时间
void invalidate()	使当前的 Session 失效，此时 Servlet 容器会释放 HttpSession 对象占用的资源
void setAttribute (String name,Object value)	将一个名/值对保存在 HttpSession 对象中
Object getAttribute(String name)	根据 name 参数返回保存在 HttpSession 对象中的属性值
Enumeration getAttributeNames()	以 Enumeration 对象返回 HttpSession 对象中所有的属性名
boolean isNew()	判断是否是新创建的 Session。如果是新创建的，返回 true，否则返回 false
void setMaxInactiveInterval (int interval)	设置一个 Session 可以处于不活动状态的最大时间间隔，以秒为单位。如果超过这个时间，Session 自动失效。如果设置为负数，表示不限制 Session 处于不活动状态的时间
int getMaxInactiveInterval()	得到当前 Session 可以处于不活动状态的最大时间间隔

HttpServletRequest 接口中也提供了与 HttpSession 相关的几个常用的方法，如表 5-2 所示。

表 5-2

方 法	说 明
String getRequestSessionId()	返回与客户端请求相关的 Session 的 ID。可能与当前的会话 ID 相同，也可能不相同
boolean isRequestedSessionIdFromCookie()	会话 ID 是由 Cookie 返回的，则返回 true
boolean isRequestedSessionIdFromURL()	会话 ID 是由重写 URL 返回的，则返回 true

(续表)

方　法	说　明
boolean isRequestedSessionIdValid()	如果客户端的会话 ID 代表的是有效会话，则返回 true，否则返回 false

下面的例子会调用我们上面提供的方法，代码如示例 5-1 所示。

示例 5-1：

```
package control;

import java.io.IOException;
import java.io.PrintWriter;
import java.text.SimpleDateFormat;
import java.util.Date;
import javax.servlet.ServletException;
import javax.servlet.http.HttpServlet;
import javax.servlet.http.HttpServletRequest;
import javax.servlet.http.HttpServletResponse;
import javax.servlet.http.HttpSession;

public class GetSessionInfoServlet extends HttpServlet {

    public void doGet(HttpServletRequest request, HttpServletResponse response)
            throws ServletException, IOException {
              doPost( request, response);
    }

    public void doPost(HttpServletRequest request, HttpServletResponse response)
            throws ServletException, IOException {
        request.setCharacterEncoding("UTF-8");
        response.setContentType("text/html;charset=UTF-8");
        PrintWriter out = response.getWriter();
        //获取与当前请求相关的会话
        HttpSession session = request.getSession();
        Date creationTime = new Date(session.getCreationTime());
        Date lastAccessed = new Date(session.getLastAccessedTime());
        Date now = new Date();
        SimpleDateFormat fmt = new SimpleDateFormat("yyyy-MM-dd HH:mm:ss");
        out.println("<html>");
        out.println("<head>");
        out.println("<title>显示会话的信息</title>");
        out.println("</head>");
        out.println("<body>");
        out.println("<h2>会话创建和最后处理时间</h2>");
        out.println("现在时间是: " + fmt.format(now) + "<br><br>");
        out.println("会话创建时间是:"+ fmt.format(creationTime) + "<br><br>");
```

```
            out.println("会话 ID 是:"+ session.getId() + "<br><br>");
            out.println("会话最大活动期是:"+ session.getMaxInactiveInterval() + "秒<br><br>");
            out.println("最后处理时间是:"+ fmt.format(lastAccessed));
            out.println("<h2>HttpRequest 的信息</h2>");
            out.println("来自请求会话 ID:" + request.getRequestedSessionId()+"<BR>");
            out.println("会话是来自 Cookie 吗？ :" + request.isRequestedSessionIdFromCookie()+"<BR>");
            out.println("会话是来自 URL 吗?:" + request.isRequestedSessionIdFromURL()+"<BR>");
            out.println("会话是有效的吗?" +request.isRequestedSessionIdValid()+"<BR>");
            //销毁会话
            session.invalidate();
            out.println("</body>");
            out.println("</html>");
        }
}
```

运行程序，显示的界面如图 5-1 所示。

图 5-1

5.3 会话跟踪

会话跟踪允许服务器确定访问站点的用户、用户访问站点的次数和用户停留站点的时间段。在客户端和服务器之间的会话 ID 和状态信息交换，Web 服务器通常有以下 4 种会话跟踪方法。

- 使用 Servlet API 中的 Session 机制。
- 使用 Cookie。
- 分别将会话 ID 存放在 URL 路径中、执行 URL 重写。
- 隐藏表单域。

5.3.1　使用 Session 的会话跟踪

我们来写一个较复杂的 Servlet 来说明会话跟踪的原理。我们知道，当客户端打开浏览器，向服务器请求某个页面的时候，服务器会为这个客户端创建一个会话对象，并用唯一的 ID 来标识容器中的会话对象。然后，服务器完成客户端的请求处理，会把与这个客户对应的会话的 ID 放在响应头中，传回给客户端，这个会话 ID 将以文件的形式保存在客户端的机器中。如果同一个客户端再次向服务器发送请求，这个会话的 ID 将被放到请求头中传给服务器，服务器使用这个 ID 和容器中的会话 ID 比较，如果是一样的，服务器就知道了这个请求和上个请求来自同一个客户端，从而就唯一地标识了客户端。服务器端的每个会话对象都有有效期，默认是 1800 秒，客户端的 Cookie 也有默认的保存时间，浏览器关闭时，Cookie 会被自动删除。所谓的会话跟踪，必须是在会话的有效期内，过了有效期，服务器就会销毁容器中的会话。客户端关闭了浏览器，Cookie 也就被自动删除了，这时，也就谈不上会话跟踪了。

本例中，这个 Servlet 提示用户选择背景色和字体并输入用户名，然后这个 Servlet 通过会话来记忆每个用户的名字、喜欢的颜色和字体及这个用户请求 Servlet 的次数。代码如示例 5-2 所示。

示例 5-2：

```
package control;

import java.io.IOException;
import java.io.PrintWriter;

import javax.servlet.ServletException;
import javax.servlet.http.HttpServlet;
import javax.servlet.http.HttpServletRequest;
import javax.servlet.http.HttpServletResponse;
import javax.servlet.http.HttpSession;

public class ColorServlet extends HttpServlet {


    public void doGet(HttpServletRequest request, HttpServletResponse response)
            throws ServletException, IOException {
        doPost(request, response);
    }

    public void doPost(HttpServletRequest request, HttpServletResponse response)
            throws ServletException, IOException {
        request.setCharacterEncoding("UTF-8");
        response.setContentType("text/html;charset=UTF-8");
        String name = "";
```

```
String color = "";
String sfont = "";
Integer hitCount = new Integer(0);
HttpSession session = request.getSession(true);
PrintWriter out = response.getWriter();
out.println("<HTML>");
out.println("<HEAD><TITLE>会话跟踪</TITLE></HEAD>");
if (session.isNew()) { //第一次请求页面，创建 Session 对象
hitCount = new Integer(1);
out.println("<BODY>");
out.print("<form action=ColorServlet method=POST>");
out.print("<H3>请选择你喜欢的背景色和字体</H3>");
out.println("<input type=radio name=bcolor value=white checked>白色 ");
out.println("<input type=radio name=bcolor value=red>红色 ");
out.println("<input type=radio name=bcolor value=green>绿色 ");
out.println("<input type=radio name=bcolor value=blue>蓝色 <br>");
out.println("<input type=radio name=sfont value=华文隶书 checked>华文隶书 ");
out.println("<input type=radio name=sfont value=宋体>宋体 ");
out.println("<input type=radio name=sfont value=方正舒体>方正舒体<br>");
out.println("请输入你的名字<br>");
out.println("<input type=text name=name value=''><br>");
out.println("<input type=submit name=submit value=提交><br>");
out.println("</form>");
} else { //第二请求，不再创建 Session 对象
        //从 request 对象中拿值
    color = request.getParameter("bcolor");
    name = request.getParameter("name");
    sfont = request.getParameter("sfont");
    //其他的更多次请求，此时，request 对象中不再有值了
    if (color != null&& name !=null &&sfont !=null) {
        hitCount = new Integer(1);
        session.setAttribute("color", color);
        session.setAttribute("hitCount", hitCount);
        session.setAttribute("name", name);
        session.setAttribute("sfont", sfont);
    } else { //从不同的 Session 对象中取值
        color = (String) session.getAttribute("color");
        name = (String) session.getAttribute("name");
        hitCount = (Integer)session.getAttribute("hitCount");
        sfont = (String)session.getAttribute("sfont");
    }
    session.setAttribute("hitCount", new Integer(hitCount.intValue() + 1));
    out.println("<BODY bgcolor=" + color + ">");
out.println("<h2><font face ="+sfont+">你好！ " + name + "</font></h2>");
    out.println("<p><font face ="+sfont+">你请求这个页面" + hitCount+ "次</font></p>");
    out.println("<p><a href=ColorServlet>请求本页面</a></p>");
}
```

```
            out.println("</BODY>");
            out.println("</HTML>");
            out.flush();
            out.close();
        }
    }
```

部署 Servlet 到 Tomcat 服务器，打开浏览器，请求该 Servlet，显示结果如图 5-2 所示。

图 5-2

当用户选择绿色和隶书并输入名字，如“张三丰”，单击“提交”按钮时，页面的背景色就会变成用户选择的颜色，字体变成隶书，显示结果如图 5-3 所示。如果用户持续单击超链接，则请求页面的次数会增加，显示结果如图 5-4 所示。

图 5-3

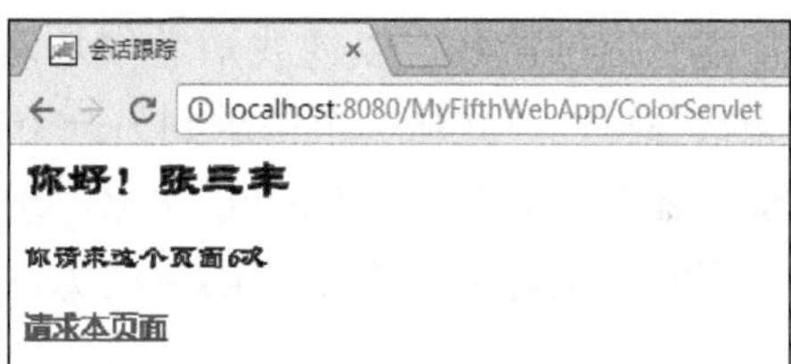

图 5-4

此时，重新打开一个浏览器，请求同一个 Servlet，将依然显示图 5-2 所示的页面，此时选择“红色”和“方正舒体”字体，输入“张无忌”，单击“提交”按钮，页面的背景色变成红色，字体变成方正舒体，显示结果如图 5-5 所示。如果用户“张无忌”持续单击“请求本页面”的超链接，页面的次数会不停地增加，结果如图 5-6 所示。打开两个不同的浏览器，请求服务器的同一个资源，相当于两个不同的客户端，服务器在每个客户端第一次请求页面的时候，会给该请求分别创建各自的会话对象，并把该会话对象放在容器中，使用唯一的 ID 来区分它们，服务器处理完请求后，ID 又被服务器放在响应头中传回到各个客户端，写入到客户端的 Cookie 中。当“张三丰”和“张无忌”用户分别单击各自页面的超链接时，浏览器又把放在 Cookie 中的会话 ID 传回给服务器，服务器就能准确地分辨出“张三丰”和“张无忌”这两个客户端喜欢的颜色和字体以及它们在页面上单击的次数，这就是会话跟踪的原理。

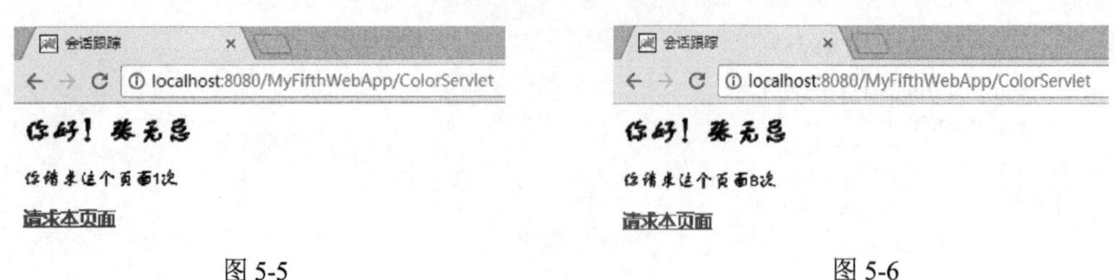

图 5-5　　　　图 5-6

5.3.2　使用 Cookie

从上面的例子知道，会话跟踪的实现是客户端和服务器端共同完成的，服务器端使用 HttpSession 接口来实现，客户端使用 Cookie 类来实现。Cookie 的本意是“小甜饼”，我们俗称“曲奇饼”，在 Java Web 中，Cookie 是 HTTP 客户端和 HTTP 服务器之间传送的小块信息，可以用这类信息将状态添加到无状态的 HTTP 协议中。当 HTTP 服务器收到一个请求时，除了被请求的文档外，服务器还会选择返回一些状态信息给客户的浏览器，而这些状态信息应该由支持 Cookie 的客户端存储起来，放在客户端的机器中。客户端任何时候再发送新的请求给服务器时，都要首先检查请求的 URL 和所存的 Cookies 中的 URL 是否对应，如果找到了，就把以前从服务器接收到的状态信息包含到客户的请求中。这就相当于服务器告诉客户这样一些事，“这是用来标识你的数字，无论你什么时候发送请求都要把它返回给我，以便我知道你是谁”。用这种方法，服务器就可以克服 HTTP 协议的无状态特性，并从一个请求到另一个请求能够实现客户追踪。

Servlet API 中提供了 Cookie 类来实现 Cookie，并定义了添加 Cookie 和获取 Cookie 的方法，同时，也定义了管理 Cookie 的方法。表 5-3 列出了 Cookie 类中的常用方法。

表 5-3

方　法	说　明
void setMaxAge(int expiry)	设置 Cookie 的最大保存时间，单位为秒，正值表示经过设定的秒值后，Cookie 将过期。负值表示不在客户机上保存 Cookie，如果用户关闭浏览器，Cookie 会被删除，Cookie 的默认设置为-1。0 值表示删除现有的 Cookie
int getMaxAge()	返回以秒表示的 Cookie 的最大保存时间
String getName()	返回在构造方法中为 Cookie 设置的名字
void setValue(String newVal)	给 Cookie 赋值，设置名/值对中值的部分
String getValue()	返回 Cookie 名/值对中值的部分

如何在 HTTP 请求和相应标题中传送 Cookie 呢？我们从两个不同的方向来看，先从服务器的响应开始。服务器在 HTTP 响应标题中设置 Cookie 的语法是：Set-Cookie:NAME=VALUE;expires=DATE;domain=DOMAIN_NAME;Path=PATH。Set-Cookie 指令被包含在服务器的 HTTP 响应的标题中，以指示客户端在本地机上保存 Cookie，也可

以把多个 Set-Cookie 标题包括在一个 HTTP 响应中。当支持 Cookie 的浏览器在服务器的响应中遇到 Set-Cookie 标题时，它就把这个 Cookie 保存起来。我们来看一个典型的 HTTP 响应标题，该响应设置了客户 Cookie。

```
HTTP/1.0   200   OK
Server :Netscape-Enterprise/2.01
Content-Type :text/html
Content-Length: 80
Set-Cookie:customerID=1234;expires=Thursday 23-Aug-2018 23:59:59 GMT;
    domain=.sina.com
Path=/order
```

任何时候客户向.sina.com 域的/order 路径请求，服务器将返回给客户端按照上面这个 HTTP 标题设置的 Cookie。例如，当使用 URL“http://www.sina.com/order/checkout.html”向新浪服务器请求 checkout.html 页面时，服务器将向客户端返回按照上面 HTTP 标题设置的 Cookie。

我们已经知道了服务器如何把 Cookie 存放在客户的机器上，接下来看看 Cookie 的信息是如何被发送到服务器的。与服务器设置 Cookie 的方式相同，客户也是在 HTTP 的请求标题中发送 Cookie 信息的。不同的是，它不是使用 Set-Cookie。客户端设置 Cookie 标题的语法是：Cookie:NAME1=VALUE1;NAME2=VALUE2;NAME3=VALUE3…任何时候客户发送请求，浏览器都试着首先将这个请求域或路径的信息与客户机上所保存的 Cookie 进行匹配，如果找到了匹配的路径和域，Cookie 的名/值对就被添加到 Cookie 的标题中。下面是一个典型的客户请求中包含的 Cookie 信息。

```
GET /login.html   http:/1.1
User-Agent:Mozilla/6.0
Accept:image/gif,image/jpeg,*/*
Cookie:customerID=1234;color=blue
```

前面的章节里学习过，Servlet API 中的 HttpServletResponse 接口提供了 setHeader()方法，可以用来把 HTTP 标题添加到响应中。同样，Servlet API 中也提供了 Cookie 类，为发送 Cookie 到客户端提供了方法。下面的代码片段说明了如何构造一个 Cookie 对象。

```
Cookie myCookie=new Cookie("customerID ","1234");
myCookie.setDomain("");
myCookie.setPath("/");
```

下面的例子演示服务器把两个 Cookie 返回给客户端，并从客户端的本地机上读出这两个 Cookie 和返回请求头中的信息。代码如示例 5-3 所示。

示例 5-3：

```
package control;

import java.io.IOException;
```

```
import java.io.PrintWriter;
import java.util.Enumeration;

import javax.servlet.ServletException;
import javax.servlet.http.Cookie;
import javax.servlet.http.HttpServlet;
import javax.servlet.http.HttpServletRequest;
import javax.servlet.http.HttpServletResponse;
import javax.servlet.http.HttpSession;

public class CookieWriteAndReadServlet extends HttpServlet {

    public void doGet(HttpServletRequest request, HttpServletResponse response)
            throws ServletException, IOException {
        doPost(request,response);
    }

    public void doPost(HttpServletRequest request, HttpServletResponse response)
            throws ServletException, IOException {
        request.setCharacterEncoding("UTF-8");
        response.setContentType("text/html;charset=UTF-8");
        //创建会话对象
        HttpSession session = request.getSession();
            //构造两个 Cookie 对象
        Cookie customerID = new Cookie("customerID", "123456");
        Cookie color = new Cookie("color", "blue");
        customerID.setMaxAge(300);
        color.setMaxAge(300);
        //把 Cookie 添加到响应头
        response.addCookie(customerID);
        response.addCookie(color);
        PrintWriter out = response.getWriter();
        out.println("<HEAD><TITLE>读取客户端的 Cookie</TITLE></HEAD>");
        out.println("<BODY>");
        out.print("客户端本机上的 cookie<table border='1'>");
        out.print("<tr><td>cookie 名字</td><td>cookie 的值</td></tr>");
          //获取客户端放在 HTTP 请求里面的所有 Cookie
        Cookie[] cookies = request.getCookies();
        if (cookies != null) {
        //以表格的形式输出 Cookie 的名字和值
            for (int i = 0; i < cookies.length; i++) {
                out.println("<tr><td>" + cookies[i].getName() + "</td>");
                out.println("<td>" + cookies[i].getValue() + "</td></tr>");
            }
        }
        out.println("</table>下面是 HTTP 请求的信息");
        out.println("<table border='1'><tr><td>头名字</td><td>头的值</td></tr>");
```

```
            //获取所有的请求头的信息
            Enumeration en = request.getHeaderNames();
            while (en.hasMoreElements()) {
            //以表格的形式输出请求信息的名字和值
                String name = (String) en.nextElement();
                out.println("<tr><td>" + name + "</td><td>"
                        + request.getHeader(name) + "</tr></td>");
            }
            out.println("</table></BODY>");
            out.println("</HTML>");
            out.flush();
            out.close();
        }
    }
```

程序运行结果如图 5-7 所示。

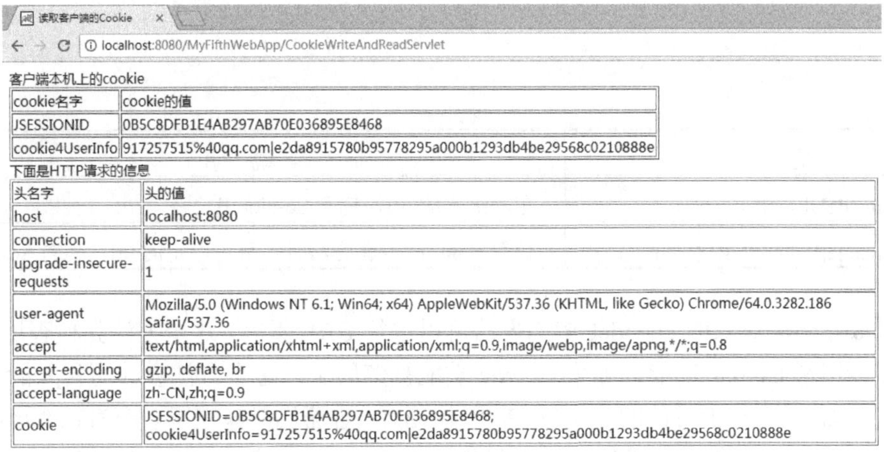

读取客户端的Cookie

localhost:8080/MyFifthWebApp/CookieWriteAndReadServlet

客户端本机上的cookie

<table>
<tr><td>cookie名字</td><td>cookie的值</td></tr>
<tr><td>JSESSIONID</td><td>0B5C8DFB1E4AB297AB70E036895E8468</td></tr>
<tr><td>cookie4UserInfo</td><td>917257515%40qq.com|e2da8915780b95778295a000b1293db4be29568c0210888e</td></tr>
</table>

下面是HTTP请求的信息

<table>
<tr><td>头名字</td><td>头的值</td></tr>
<tr><td>host</td><td>localhost:8080</td></tr>
<tr><td>connection</td><td>keep-alive</td></tr>
<tr><td>upgrade-insecure-requests</td><td>1</td></tr>
<tr><td>user-agent</td><td>Mozilla/5.0 (Windows NT 6.1; Win64; x64) AppleWebKit/537.36 (KHTML, like Gecko) Chrome/64.0.3282.186 Safari/537.36</td></tr>
<tr><td>accept</td><td>text/html,application/xhtml+xml,application/xml;q=0.9,image/webp,image/apng,*/*;q=0.8</td></tr>
<tr><td>accept-encoding</td><td>gzip, deflate, br</td></tr>
<tr><td>accept-language</td><td>zh-CN,zh;q=0.9</td></tr>
<tr><td>cookie</td><td>JSESSIONID=0B5C8DFB1E4AB297AB70E036895E8468;
cookie4UserInfo=917257515%40qq.com|e2da8915780b95778295a000b1293db4be29568c0210888e</td></tr>
</table>

图 5-7

5.3.3　URL 重写

从上面的分析我们知道，在 Java Web 中，使用会话和 Cookie 来跟踪客户。Cookie 存放在客户端的机器上面，如果客户端的浏览器出于安全方面的考虑，在浏览器的隐私设置中阻止了所有的 Cookie(见图 5-8)，这样，服务器就没有办法把一些信息写入到客户端，使用 Cookie 就没有用了。这个时候，我们还是需要跟踪客户，那么怎样才能跟踪到客户呢？Java Servlet API 中为我们提供了会话跟踪的另一种机制，通常的做法是重写客户请求的 URL，把 Session ID 添加到 URL 信息中。

URL 重写其实就是在 URL 后面加上一个 jsessionid 参数，jsessionid 的参数值是唯一的，所以可以跟踪某一会话。HttpServletResponse 接口中提供了两种 URL 重写的方法，如表 5-4 所示。

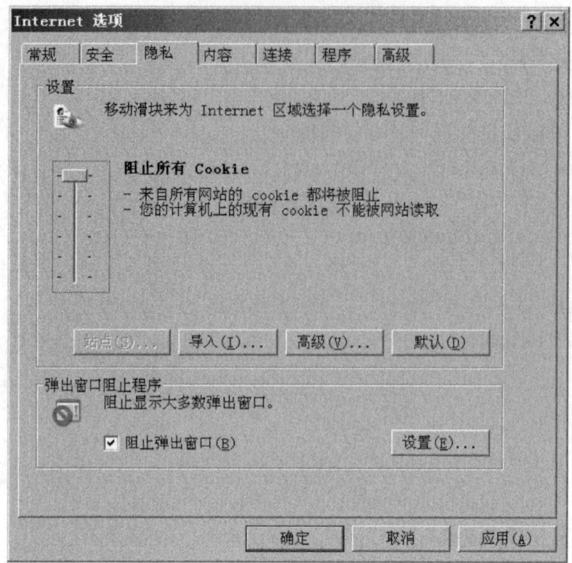

图 5-8

表 5-4

方　法	说　明
String encodeURL(String url)	重写给定的 url，包含 Session ID
String encodeRedirectURL(String url)	使用 sendRedirect 方法时，重写给定的 url，包含 Session ID

通过 response 对象的 encodeURL()方法可以进行 URL 重写，这个方法先检查用户端浏览器是否禁用了 Cookie 功能，如果 Cookie 功能被禁用了，这个方法会改写传入的 URL，方式是在 URL 后面加上目前的 jsessionid(假设 Servlet 容器已经建立了一个 HTTP 会话)。如果 Cookie 功能并未被禁用，或是目前并未建立 HTTP 会话，这个方法会保留传入的 URL，不做任何改动。很明显，URL 重写并不依赖于 Cookie，而是依赖于 URL 参数，类似于 http://.../result.jsp?jsessionid=xxxxxxxxxx 这样的形式，Web 服务器通过获取 jsessionid 的值来找到对应的 HttpSession 对象，然后进行操作。而使用 URL 重写的麻烦就是，需要对每一个需要跟踪的会话的 URL 进行 URL 重写。

来看一个例子，本例中，假定客户端更改了浏览器的安全设置，阻止了所有的 Cookie，看看请求的 URL 发生了什么变化。建立一个名叫 UrlRewriteServlet 的 Servlet，代码如示例 5-4 所示。在该页面中，有一个链接到 default.html 页面的超链接，代码如示例 5-5 所示。

示例 5-4：

```
package control;

import java.io.IOException;
import java.io.PrintWriter;
import javax.servlet.ServletException;
import javax.servlet.http.HttpServlet;
```

```
import javax.servlet.http.HttpServletRequest;
import javax.servlet.http.HttpServletResponse;
import javax.servlet.http.HttpSession;

public class UrlRewriteServlet extends HttpServlet {

    public void doGet(HttpServletRequest request, HttpServletResponse response)
            throws ServletException, IOException {
        doPost(request,response);
    }

    public void doPost(HttpServletRequest request, HttpServletResponse response)
            throws ServletException, IOException {
        request.setCharacterEncoding("UTF-8");
        response.setContentType("text/html;charset=UTF-8");
        PrintWriter out = response.getWriter();
        //创建会话对象
        HttpSession session=request.getSession();
        //获取路径
        String contextPath = request.getContextPath();
        //使用 URL 重写方法
        String newURL = response.encodeURL(contextPath + "/default.html");
        out.println("<HTML>");
        out.println("<HEAD><TITLE>演示客户端禁用浏览器的 Cookie</TITLE></HEAD>");
        out.println("<BODY>");
        //根据新的 URL 地址来判断客户端是否禁用 Cookie
        if(newURL.endsWith(session.getId())){
            out.println("客户端禁用了 Cookie，URL 将重写<br>");
            out.println("客户请求的 default.hmtl 页面新的 URL 是:<br>"+newURL+"<br>");
        }
        else{
            out.println("客户端使用了 Cookie，URL 将不变<br>");
            out.println("客户请求的 default.hmtl 页面 URL 是:<br>"+newURL+"<br>");
        }
        out.println("<a href="+ newURL+ ">去到 default.html</a>");
        out.println("</BODY>");
        out.println("</HTML>");
        out.flush();
        out.close();
    }
}
```

示例 5-5：

```
<html>
  <head>
    <title>演示 URL 重写</title>
```

```
    </head>
   <body>
       <h1>使用了 URL 重写</h1>
  </body>
  </html>
```

部署该 Servlet 应用，启动 Tomcat 服务器，打开 IE 浏览器，在 IE 的 Internet 选项中禁用 Cookie，在地址栏输入 http://1ocalhost:8080/MyFifthWebApp/UrlRewriteServlet，结果如图 5-9 所示。并且页面记录了客户请求 default.html 页面的 URL 地址为：/MyFifthWebApp/default.html;jsessionid=EB66BD621A3B0432C70855B3F7DD57FD。

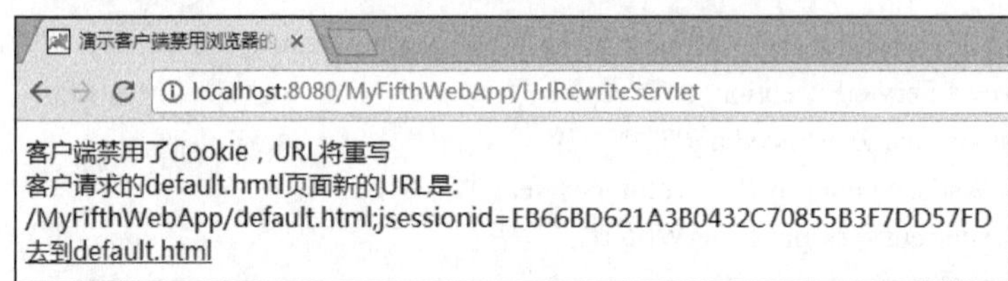

图 5-9

单击“去到 default.html”超链接，将显示 default.html 的页面，IE 地址栏中的路径内容与图 5-9 中显示的 URL 路径内容是一样的，因为客户端的浏览器禁用了 Cookie，所以容器就决定重写客户请求的 URL，把 jsessionid=xxx 附加到客户请求的 URL 后面，目的是标识客户端，完成客户端的跟踪，如图 5-10 所示。

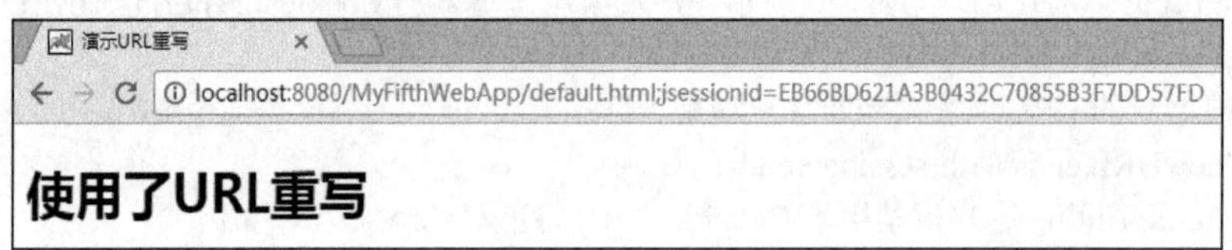

图 5-10

此时，如果打开对 Cookie 的支持，同样来请求 MyFifthWebApp 应用下的 UrlRewriteServlet，结果如图 5-11 所示。

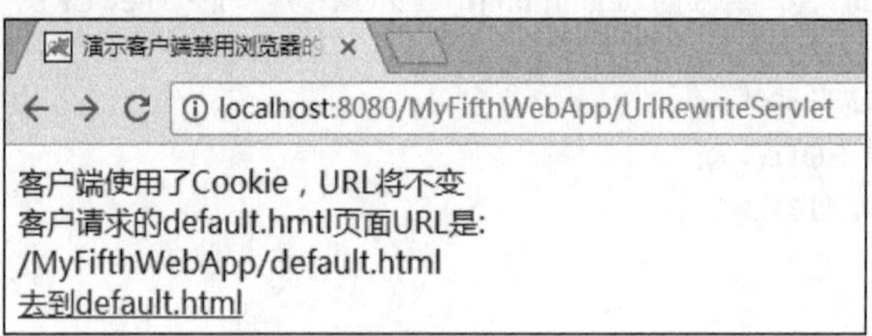

图 5-11

此时，单击“去到 default.html”超链接，将显示 default.html 的页面，IE 地址栏中的内容与图 5-11 中页面显示的客户请求的 default.html 页面 URL 路径内容是一样的，容器并没有重写客户请求的 URL 地址，如图 5-12 所示。

图 5-12

HttpServletResponse 接口中还有另外一个 URL 重写方法：encodeRedirectURL(String url)，当用户使用 response.sendRedirect(String url)来重定向时，如果要重定向的目的地客户同样禁用了浏览器的 Cookie 功能，也必须要使用 URL 重写。我们在后续的章节里会介绍 sendRedirect()方法的使用。

5.3.4　隐藏表单域

隐藏 HTML 表单域通常用来存储状态信息。隐藏的变量操作就像 HTML 输入域(文本输入域、复选框和单选按钮)一样，提交页面给服务器时，客户端会把该隐藏域的名/值对传给服务器。不同之处在于，客户端的页面不能看到或是修改隐藏的 HTML 变量的值，但是在 HTML 的源代码中可以看到变量的值。

下面的代码片段显示了在 HTML 表单中隐藏表单域的用法。

```
<form action="registerServlet" method = "post">
 邮件地址：<input type="text" name = "email" ><br>
 年　龄 ：<input type="text" name = "age"><br>
 <!--使用隐藏表单域来传递两个数据-->
 <input type="hidden" name="uname" value="张三丰">
 <input type="hidden" name="pwd" value="998">
 <input type= "submit" value = "提交到服务器">
</form>
```

【单元小结】

- Servlet API 中使用 HttpSession 和 Cookie 来跟踪客户的状态。
- getSession()方法返回与客户端请求绑定的会话对象。
- 会话跟踪的 4 大技术分别是使用 Session 和 Cookie、URL 重写、隐藏表单域。

【单元自测】

1. 下列哪个方法能得到与客户端请求绑定的会话对象？(　　)

   A. getSession()　　　　　　B. new HttpSession()

   C. new SessionInstence()　　　D. getSession(false)

2. 下列不属于会话跟踪技术内容的是(　　)。

A. URL 重写　　B. Cookie

C. 隐藏表单域　　D. 查询字符串

3. 下面不属于 Cookie 头部内容的是(　　)。

A. name　　B. expires

C. domain　　D. value

4. Cookie 方法中，哪个能得到 Cookie 的存在时间？(　　)

A. getMaxAge()　　B. setMaxAge()

C. getValue()　　D. getName()

【上机实战】

上机目标

- 了解带参数的 Servlet。
- 熟练使用查询字符串。

上机练习

◆ 第一阶段 ◆

练习 1：练习使用带参数的 Servlet

【问题分析】

(1) 实际的 Web 应用中，可以把参数附加在请求 URL 后面，然后把这些参数值传给页面或 Servlet。通常采用两种方式，一种是放在表单的 action 属性里面，一种是放在超链接 href 属性里面。

(2) 使用 request 对象的 getParameter 方法能够获取参数的值。

【参考步骤】

(1) 创建 Web 工程，取名为 SessionTrackDemo1。

(2) 修改工程中的 index.jsp，如示例 5-6 所示。

示例 5-6：

```
<%@ page language="java"    pageEncoding="UTF-8"%>
<html>
    <head>
        <title>登录页面</title>
    </head>
```

```
    <body>
        <h3>欢迎登录</h3>
        <form action="ParameterServletServlet?age=18&sex=male&country=china"
method="post">
            请输入你的登录名：
            <input type="text" name="username" value="">
            <br>
            请输入你的密码：
            <input type="password" name="userpwd" value="">
            <br>
            <input type="submit" name="submit" value="提交">
        </form>
        <br>
        <a href = "ParameterServlet?age=18&sex=male&country=china">去到 Servlet</a>
    </body>
</html>
```

(3) 创建一个名叫 ParameterServlet 的 Servlet，代码如示例 5-7 所示。

示例 5-7：

```
package control;

import java.io.IOException;
import java.io.PrintWriter;

import javax.servlet.ServletException;
import javax.servlet.http.HttpServlet;
import javax.servlet.http.HttpServletRequest;
import javax.servlet.http.HttpServletResponse;

public class ParameterServlet extends HttpServlet {

    public void doGet(HttpServletRequest request, HttpServletResponse response)
            throws ServletException, IOException {
                doPost(request,response);
    }

    public void doPost(HttpServletRequest request, HttpServletResponse response)
            throws ServletException, IOException {
        response.setContentType("text/html;charset=UTF-8");
        PrintWriter out = response.getWriter();
        out.println("<HEAD><TITLE>显示信息</TITLE></HEAD>");
        out.println("<BODY>");
        String name = request.getParameter("username");
        String pwd = request.getParameter("userpwd");
        String age = request.getParameter("age");
        String sex = request.getParameter("sex");
```

```
            String country = request.getParameter("country");
            out.print("个人信息如下:<br>");
            out.print("用户名:"+name+"<br>");
            out.print("密码:"+pwd+"<br>");
            out.print("年龄:"+age+"<br>");
            out.print("性别:"+sex+"<br>");
            out.print("国籍:"+country+"<br>");
            out.println("</BODY>");
            out.println("</HTML>");
            out.flush();
            out.close();
        }
    }
```

(4) 发布该应用到 Tomcat 服务器并启动，打开 IE 浏览器并在地址栏中输入“http://localhost:8080/SessionTraceDemo1/”，输出结果如图 5-13 所示。

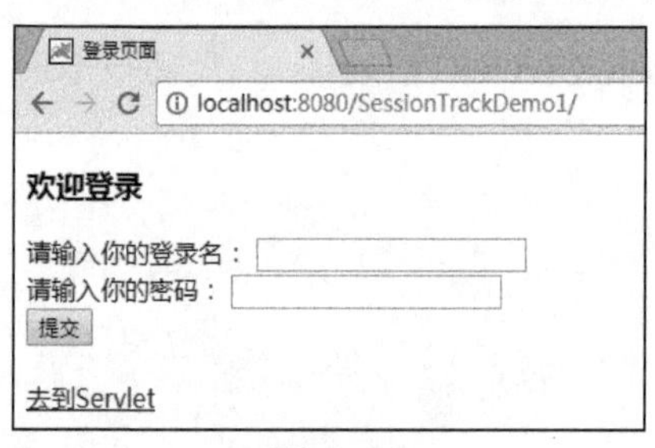

图 5-13

(5) 输入用户名和密码，如“bush”和“888”，单击“提交”按钮，显示如图 5-14 所示的界面。

(6) 单击“去到 Servlet”超链接，显示如图 5-15 所示的界面。

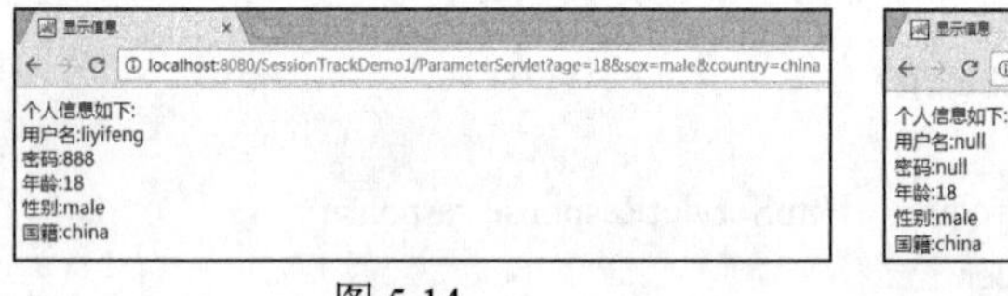

图 5-14

图 5-15

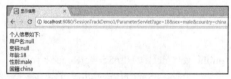

(7) 从图中可以看出，请求 URL 后面附加的参数值在两种方式下都传给了请求的 Servlet。想一想，为什么表单中的数据在超链接的方式下不能被传给请求的 Servlet?

◆ 第二阶段 ◆

练习 2：练习使用带参数的 Servlet

【问题描述】

数据库中有一个 books 表，存放了书籍的相关信息，如 ISBN 编号、书名、作者、价格和库存数量等信息。当我们新进了同样的书时，希望能更改库存数量。同时，如果有折扣，我们同样希望可以更改书的价格。

【问题分析】

(1) 使用三层结构完成数据库操作，使用业务类完成数据库 books 表的字段的更新操作。

(2) 需要建立一个 Servlet，用来以表格的形式显示数据库中 book 表的所有信息，同时提供超链接。

(3) 需要建立另外的 Servlet，完成数据库 books 表的字段的修改。

(4) Servlet 后面可以跟键/值对的字符串，多个键/值对可以使用“&”符号隔开。

【参考步骤】

(1) 创建 Web 工程，取名为 SessionTraceDemo2。

(2) 建立数据库操作支持类 DBHelper.java、DBCommand.java 类、数据库属性配置文件。

(3) 新建与数据库表 books 对应的实体类 BookBean.java，代码如示例 5-8 所示。

示例 5-8：

```
package bean;

public class BookBean {

    private Integer bookId;
    private String isbn;
    private String name ;
    private String author;
    private Float price ;
    private Integer stock;

    public String getIsbn() {
        return isbn;
    }
    public void setIsbn(String isbn) {
        this.isbn = isbn;
    }
    public String getName() {
        return name;
    }
    public void setName(String name) {
        this.name = name;
    }
    public String getAuthor() {
        return author;
    }
    public void setAuthor(String author) {
```

```
        this.author = author;
    }
    public Integer getBookId() {
        return bookId;
    }
    public void setBookId(Integer bookId) {
        this.bookId = bookId;
    }
    public Float getPrice() {
        return price;
    }
    public void setPrice(Float price) {
        this.price = price;
    }
    public Integer getStock() {
        return stock;
    }
    public void setStock(Integer stock) {
        this.stock = stock;
    }
```

(4) 建立操作数据库 Books 表的基础类，BookDao.java 和 BookDAOImpl.java 的代码分别如示例 5-9、示例 5-10 所示。

示例 5-9：

```
package dao;

import java.util.List;
import bean.BookBean;

public abstract class BookDAO extends BaseDAO {
    public abstract List<BookBean> getAllBooks() ;
    public abstract int updateBook(BookBean bookBean);
    public abstract BookBean getBookById(Integer id);
}
```

示例 5-10：

```
package dao;

import java.sql.*;
import java.util.List;
import java.util.ArrayList;
import java.util.Map;
import java.util.Iterator;
import java.util.LinkedHashMap;
```

```
import support.DBCommand;
import bean.BookBean;

public class BookDAOImpl extends BookDAO {
    private PreparedStatement pstm;
    // 获取所有的图书
    public List<BookBean> getAllBooks() {
        List<BookBean> bookList = new ArrayList<BookBean>();
        try {
            pstm = this.getConn().prepareStatement("SELECT * FROM books");
            List<Map<String, Object>> list = DBCommand.execQuery(pstm);
            // 迭代
            Iterator<Map<String, Object>> it = list.iterator();
            // 从 list 内提取每一行，放入 StudentEntity，最终放入 stuList
            while (it.hasNext()) {
                Map<String, Object> row = it.next();
                BookBean bookBean = new BookBean();
                bookBean.setBookId(new Integer(row.get("bookid").toString()));
                bookBean.setIsbn(row.get("isbn").toString());
                bookBean.setName(row.get("name").toString());
                bookBean.setAuthor(row.get("author").toString());
                bookBean.setPrice(new Float(row.get("price").toString()));
                bookBean.setStock(new Integer(row.get("stock").toString()));
                bookList.add(bookBean);
            }
        } catch (Exception e) {
            e.printStackTrace();
        }
        return bookList;
    }
    // 修改图书信息
    public int updateBook(BookBean bookBean) {
        try {
        pstm = this.getConn().prepareStatement("update books set
            isbn=?,name=?,author=?,price=?,stock = ? where bookid=?");
        Map<Object, Object> paramsMap = new LinkedHashMap<Object, Object>();
            paramsMap.put("isbn", bookBean.getIsbn());
            paramsMap.put("name", bookBean.getName());
            paramsMap.put("author", bookBean.getAuthor());
            paramsMap.put("price", bookBean.getPrice());
            paramsMap.put("stock", bookBean.getStock());
            paramsMap.put("bookid", bookBean.getBookId());
            return DBCommand.execUpdate(pstm, paramsMap);
        } catch (Exception e) {
            e.printStackTrace();
        }
        return -1;
```

```
    }

    // 根据 bookid 查找图书
    public BookBean getBookById(Integer id) {
        BookBean bookBean = null;
        try {
        pstm = this.getConn().prepareStatement(
                    "SELECT * FROM books WHERE bookid = ?");
        Map<Object, Object> paramsMap = new LinkedHashMap<Object, Object>();
            paramsMap.put("bookid", id.intValue());
            List<Map<String, Object>> stuList = DBCommand.execQuery(pstm,
                    paramsMap);
            if (stuList.size() != 0) {
                bookBean = new BookBean();
                Map<String, Object> row = stuList.get(0);
                bookBean.setBookId(new Integer(row.get("bookid").toString()));
                bookBean.setIsbn(row.get("isbn").toString());
                bookBean.setName(row.get("name").toString());
                bookBean.setAuthor(row.get("author").toString());
                bookBean.setPrice(new Float(row.get("price").toString()));
                bookBean.setStock(new Integer(row.get("stock").toString()));
            }
        } catch (Exception e) {
            e.printStackTrace();
        }
        return bookBean;
    }
}
```

(5) 建立操作数据库 Books 表的业务类：BookService.java 和 BookServiceImpl.java，代码分别如示例 5-11、示例 5-12 所示。

示例 5-11：

```
package service;

import java.util.List;
import bean.BookBean;

public interface BookService{
    public abstract List<BookBean> getAllBooks() ;
    public abstract boolean updateBook(BookBean bookBean);
    public abstract BookBean getBookById(Integer id);
}
```

示例 5-12：

```
package service;
```

```
import java.util.List;
import bean.BookBean;
import dao.BookDAO;
import dao.BookDAOImpl;
import support.DBHelper;

public class BookServiceImpl implements BookService {

    private BookDAO dao = new BookDAOImpl();

    public BookBean getBookById(Integer id) {
        BookBean book = null;
        try {
            dao.setConn(DBHelper.getConn());
            book= dao.getBookById(id);
        } catch (RuntimeException e) {
            e.printStackTrace();
        }finally{
            DBHelper.closeConn();
        }
        return book;
    }

    public List<BookBean> getAllBooks() {
        List<BookBean> bookList = null;
        try {
            dao.setConn(DBHelper.getConn());
            bookList = dao.getAllBooks();
        } catch (RuntimeException e) {
            e.printStackTrace();
        }finally{
            DBHelper.closeConn();
        }
        return bookList;
    }

    public boolean updateBook(BookBean bookBean) {
         boolean res = false;
         try {
            dao.setConn(DBHelper.getConn());
             res = dao.updateBook(bookBean) >0 ? true :false ;
             } catch (RuntimeException e) {
            e.printStackTrace();
        }finally{
            DBHelper.closeConn();
        }
```

```
        return res ;
    }
}
```

(6) 创建一个名叫 ShowBookServlet 的 Servlet，用来以列表的方式显示所有的书籍，如示例 5-13 所示。

示例 5-13：

```
package control;

import java.io.IOException;
import java.io.PrintWriter;
import javax.servlet.ServletException;
import javax.servlet.http.HttpServlet;
import javax.servlet.http.HttpServletRequest;
import javax.servlet.http.HttpServletResponse;
import bean.BookBean;
import java.util.List;
import service.BookService;
import service.BookServiceImpl;

public class ShowBookServlet extends HttpServlet {

    public void doGet(HttpServletRequest request, HttpServletResponse response)
            throws ServletException, IOException {
        doPost(request, response);
    }

    public void doPost(HttpServletRequest request, HttpServletResponse response)
            throws ServletException, IOException {
        request.setCharacterEncoding("UTF-8");
        response.setContentType("text/html;charset=UTF-8");
        // 实例数据库操作类
        BookService service = new   BookServiceImpl();
        List<BookBean> list = service.getAllBooks();
        PrintWriter out = response.getWriter();
        out.println("<HTML>");
        out.println("<HEAD><TITLE>取数据库数据</TITLE></HEAD>");
        out.println("<h3>图书信息如下:</h3>");
        out.println("单击图书编号可以查看详细信息");
        out.println("<BODY><table border=1>");
        out.print("<tr><td>图书编号</td><td>图书 ISBN 号</td>");
        for (int i = 0; i < list.size(); i++) {
            BookBean book = (BookBean) list.get(i);
            out.println("<tr>");
            out.println("<td><a href=showBookDetailServlet?bookid="
                    + book.getBookId() + ">" + book.getBookId() + "</a></td>");
```

```
            out.println("<td>" + book.getIsbn() + "</td>");
            out.println("</tr>");
        }
        out.println("</table></BODY>");
        out.println("</HTML>");
        out.flush();
        out.close();
    }
}
```

(7) 创建一个名叫 ShowBookDetailServlet 的 Servlet，用户单击图书编号时能链接到该 Servlet，显示书籍的详细信息，代码如示例 5-14 所示。

示例 5-14：

```
package control;

import java.io.IOException;
import java.io.PrintWriter;
import javax.servlet.ServletException;
import javax.servlet.http.HttpServlet;
import javax.servlet.http.HttpServletRequest;
import javax.servlet.http.HttpServletResponse;
import bean.BookBean;
import service.BookService;
import service.BookServiceImpl;

  @WebServlet("/ShowBookDetailServlet")
public class ShowBookDetailServlet extends HttpServlet {
  public void doGet(HttpServletRequest request, HttpServletResponse response) throws ServletException,
IOException {
      doPost(request, response);
  }

  public void doPost(HttpServletRequest request, HttpServletResponse response) throws ServletException,
IOException {
      request.setCharacterEncoding("UTF-8");
      response.setContentType("text/html;charset=UTF-8");
      String bookid = request.getParameter("bookid");
      Integer id = new Integer(bookid);
      // 构造业务对象
      BookService service = new BookServiceImpl();
      BookBean book = service.getBookById(id);
      PrintWriter out = response.getWriter();
      out.println("<HTML>");
      out.println("<HEAD><TITLE>取数据库数据</TITLE></HEAD>");
      out.println("<h3>图书详细信息如下:</h3>");
      out.println("可以修改数量和单价");
```

```
        out.println(
                " <BODY><form action=updateBookServlet?bookid=" + book.getBookId() + "
method=post><table border=1>");
        out.print("<tr><td>图书编号：</td><td>" + book.getBookId() + "</td></tr>");
        out.print("<tr><td>图书名字：</td><td>" + book.getName() + "</td></tr>");
        out.print("<tr><td>图书 ISBN 编号:</td><td>" + book.getIsbn() + "</td></tr>");
        out.print("<tr><td>图书作者：</td><td>" + book.getAuthor() + "</td></tr>");
        out.print("<tr><td>图书单价：</td><td><input type=text name=newPrice size=6 " + "value=" +
book.getPrice()
                + "></td></tr>");
        out.print("<tr><td>图书库存数量：</td><td><input type=text name=newStock size=6 " + "value="
+ book.getStock()
                + "></td></tr>");
        out.print("<tr><td colspan='2' align=center><input type=submit name=submit value=修改>" + "
</td></tr>");
        out.println("</table></form></BODY>");
        out.println("</HTML>");
        out.flush();
        out.close();
    }

    }}
```

(8) 创建一个名叫 UpdateBookServlet 的 Servlet，使用户单击“修改”按钮时能链接到该 Servlet，更改数据库中书籍的数量和价格，代码如示例 5-15 所示。

示例 5-15：

```
package control;

import java.io.IOException;
import java.io.PrintWriter;
import javax.servlet.ServletException;
import javax.servlet.http.HttpServlet;
import javax.servlet.http.HttpServletRequest;
import javax.servlet.http.HttpServletResponse;
import bean.BookBean;
import service.BookService;
import service.BookServiceImpl;

public class UpdateBookServlet extends HttpServlet {

    public void doGet(HttpServletRequest request, HttpServletResponse response)
            throws ServletException, IOException {
        doPost(request, response);
    }

    public void doPost(HttpServletRequest request,HttpServletResponse response)
```

```
            throws ServletException, IOException {
        request.setCharacterEncoding("UTF-8");
        response.setContentType("text/html;charset=UTF-8");
        //从页面来的数据
        String bookid=request.getParameter("bookid");
        String price=request.getParameter("newPrice");
        String stock=request.getParameter("newStock");
        Integer id = new Integer(bookid);
        BookService service = new BookServiceImpl();
        //根据 id 查找书籍
         BookBean book = service.getBookById(id);
         //更新书籍字段
        book.setPrice(new Float(price));
        book.setStock(new Integer(stock));
        boolean isSuccess=service.updateBook(book);
        PrintWriter out = response.getWriter();
        out.println("<HTML>");
        out.println("   <HEAD><TITLE>A Servlet</TITLE></HEAD>");
        out.println("   <BODY>");
        if(isSuccess){
        out.println("修改成功！<a href=showBookServlet>回到主页</a>");
        }
        else{
            out.println("修改失败！<a href=updateBookServlet?bookid="
                +bookid+"">回到修改页面</a>");
        }
        out.println("   </BODY>");
        out.println("</HTML>");
        out.flush();
        out.close();
    }
}
```

(9) 发布该应用到Tomcat服务器并启动，打开IE浏览器并在地址栏中输入“http://localhost:8080/SessionTraceDemo2/showBookServlet”，显示效果如图 5-16 所示。

(10) 当用户单击“图书编号”超链接时，显示效果如图 5-17 所示。

图 5-16

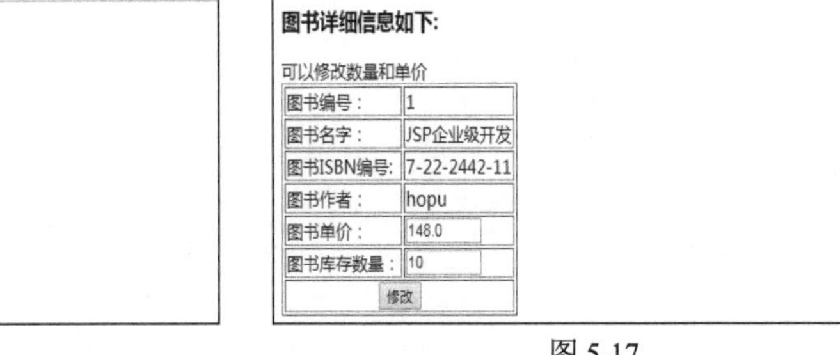

图 5-17

(11) 当用户修改图书单价或者是数量时，单击“修改”按钮，显示效果如图 5-18 所示。

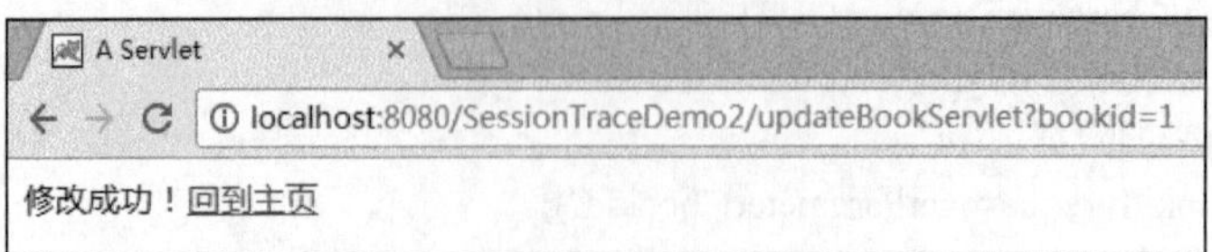

图 5-18

【拓展作业】

1. 使用 Servlet 创建一个页面，当用户登录到此页面时创建 Cookie 并存储用户登录的次数，且显示用户总的登录次数。如果该用户第 3 次登录，则为用户显示其已经成为幸运客户的信息。

2. 改进第二阶段的练习，实现删除图书的功能。以表格的形式显示图书，给每个图书增加一个“删除”按钮，单击“删除”按钮就能删除该图书。

单元六

会话和 Servlet 核心接口应用

课程目标

- 掌握会话对象的使用
- 掌握 Servlet 之间的通信问题
- 掌握 Servlet 的转发对象

简 介

前一单元我们学习了会话管理，并且知道了使用会话来跟踪客户的方式实质上都是围绕着如何处理 Session ID 的问题。这个 Session ID 实质就是客户端在请求页面时身份的标志，只要标志在，客户就不会丢失。我们想想，当在站点浏览不同页面的时候，只要不关闭浏览器，可以从一个页面导航到另外的页面，一直处于同一个会话中，这个时候，我们可以利用这个会话对象带来的便利。既然不同的页面处于同一会话中，那么我们有理由使用会话对象来达到数据共享的目的。某个页面中，在会话对象中放入数据，则在另外的页面中就能拿到放入的数据，从而实现处于同一个会话中不同页面的数据共享。

在 Servlet 体系结构中，除了用于实现 Servlet 的 Servlet 接口和 HttpServlet 类外，还有一些辅助 Servlet 获取相关资源信息的重要接口，掌握这些核心接口的常用方法将便于我们进行 Web 应用开发。

6.1 使用 Session 实现 Servlet 之间的通信

Session 对象除了用来跟踪会话外，很多时候用来实现站点间页面数据的共享。当客户端访问某个站点时，就开始了会话，此时只要客户端不关闭浏览器，它可以在站点里不同的页面中来回跳转，请求不同的页面。可以这样说，客户浏览的页面处于同一个会话中，所以可以实现数据在不同页面间的传递，从而实现页面间的通信。

6.1.1 验证用户是否登录

我们都有这样的经验，有些站点中的某些页面，如果不是登录用户，则不会显示页面的内容，一直提示要我们登录。这个功能是如何实现的呢？其实，就是利用会话实现了页面间的数据共享。首先，我们提交表单数据到验证用户的 Servlet，该 Servlet 如果验证用户是合法用户，就会在会话中设置一个标志。当我们请求其他页面时，先检查会话中有没有标志，如果没有，说明是非登录用户，把页面转到登录页面；如果会话中有标志，则显示页面的内容。

下面的例子实现了验证用户是否登录的功能，示例 6-1 是用户的登录页面，提供用户名和密码表单数据到验证 Servlet。

示例 6-1：

```
<%@ page language="java" pageEncoding="UTF-8"%>
<html>
  <head>
    <title>登录页面</title>
  </head>
```

```
    <body>
    <form action="checkServlet" method = "post">
     用户名:<input type="text" name="uname" value=""><br>
     密码:<input type="password" name="pwd" value=""><br>
     <input type= "submit" value = "登录">
    </form>
    </body>
</html>
```

CheckServlet 用来验证用户名和密码，同时，把标志信息放入会话中，代码如示例 6-2 所示。

示例 6-2：

```
package control;

import java.io.IOException;
import java.io.PrintWriter;
import javax.servlet.ServletException;
import javax.servlet.http.HttpServlet;
import javax.servlet.http.HttpServletRequest;
import javax.servlet.http.HttpServletResponse;
import javax.servlet.http.HttpSession;

public class CheckServlet extends HttpServlet {

    public void doGet(HttpServletRequest request,
    HttpServletResponse response)
            throws ServletException, IOException {
        doPost(request,response);
    }

    public void doPost(HttpServletRequest request,HttpServletResponse response)
            throws ServletException, IOException {
        response.setContentType("text/html;charset=UTF-8");
        PrintWriter out = response.getWriter();
          String name = request.getParameter("uname");
          String pwd = request.getParameter("pwd");
          //获取 Session
          HttpSession session = request.getSession();
        out.println("<HTML>");
        out.println(" <HEAD><TITLE>Check Servlet</TITLE></HEAD>");
        out.println(" <BODY>");
         //验证用户是不是合法用户
        if(name.equals("bush")&& pwd.equals("8888")){
             out.println("欢迎你！ "+name);
        //把标志放到 Session 中
            session.setAttribute("flag", name);
```

```
            }
            else{
                out.println("用户名或密码不正确");
            }
            out.println("<br><a href=showInfoServlet>去到显示信息的页面</a>");
            out.println("</BODY>");
            out.println("</HTML>");
            out.flush();
            out.close();
        }
    }
```

在 CheckServlet 中有一个超链接，可以打开显示信息的页面，这其实是一个 Servlet。在这个 Servlet 中通过判断 Session 中有没有标志来确定用户是否登录。如果没有登录，就没有对象存放到 Session 中，则提示用户“欢迎光临太平洋电脑网，你还没有登录，信息无法显示！”，如果用户已经登录，则使用表格的形式显示电脑配件的最新报价。代码如示例 6-3 所示。

示例 6-3：

```
package control;

import java.io.IOException;
import java.io.PrintWriter;
import javax.servlet.ServletException;
import javax.servlet.http.HttpServlet;
import javax.servlet.http.HttpServletRequest;
import javax.servlet.http.HttpServletResponse;
import javax.servlet.http.HttpSession;

public class ShowInfoServlet extends HttpServlet {

    public void doGet(HttpServletRequest request,
    HttpServletResponse response)
            throws ServletException, IOException {
            doPost(request,response);
    }

    public void doPost(HttpServletRequest request,
    HttpServletResponse response)
            throws ServletException, IOException {
        response.setContentType("text/html;charset=UTF-8");
        PrintWriter out = response.getWriter();
        //获取 Session
        HttpSession session = request.getSession();
        out.println("<HTML>");
        out.println("<HEAD><TITLE>A Servlet</TITLE></HEAD>");
```

```
        out.println("<BODY>");
        Object flag = session.getAttribute("flag");
        if(flag ==null){
            out.println("欢迎光临太平洋电脑网，你还没有登录，信息无法显示!<br>");
            out.println("<a href=index.jsp>回到登录页面</a>");
        }
        else{
            out.println("欢迎光临太平洋电脑网，我们提供全国最新电脑配件报价<br>");
            out.println("<table border=1><tr><td>名称</td><td>描述</td><td>价格</td></tr>");
            out.println("<tr><td>Intel Core 2 Duo E7200</td><td>接口类型:LGA 775/生产工
                艺:0.045um/主频:2.53GHz/二级缓存:L2</td><td>890 元</td></tr>");
            out.println("<tr><td>金士顿 DDR2 667 1G</td><td>内存容量:1024M/内存主频:DDR2
                667/颗粒封装:BGA</td><td>149 元</td></tr>");
            out.println("<tr><td>希捷 500G SATAII 32M</td><td>容量:500G/接口标准:S-ATA/盘体
                尺寸:3.5 寸/转速:7200rpm/缓存容量:32M</td><td>685 元</td></tr>");
            out.println("</table>");
        }
        out.println("</BODY>");
        out.println("</HTML>");
        out.flush();
        out.close();
    }
}
```

6.1.2　Servlet 间的数据共享

我们知道，用户在站点浏览页面时，只要不关闭浏览器，不同的页面就会处于同一个 Session 里，而这个 Session 默认的生存时间是 1800 秒。这就意味着，我们可以通过使用 Session 来完成页面数据的共享，从而实现页面间的数据通信功能。不同的页面在 Session 的生存期内共用同一个 Session 实例对象，只要其中的一个页面把数据保存到 Session 里面，其他的页面就能够把数据从 Session 里取出来，同时，它也可以把另外的数据存放到 Session 对象里，供其他页面使用。

下面的例子演示了两个 Servlet 使用 Session 对象来共享数据，其中 SetAttributeServlet 向 Session 中放入两个对象，一个是 HashMap 对象，另一个是 Vector 对象，第二个 Servlet 从 Session 中取出这两个对象，并把它们的值显示在页面里。代码如示例 6-4 和示例 6-5 所示。

示例 6-4：

```
package control;
import java.io.IOException;
import java.io.PrintWriter;
import java.util.HashMap;
import java.util.Vector;
```

```
import javax.servlet.ServletException;
import javax.servlet.annotation.WebServlet;
import javax.servlet.http.HttpServlet;
import javax.servlet.http.HttpServletRequest;
import javax.servlet.http.HttpServletResponse;
import javax.servlet.http.HttpSession;

@WebServlet("/SetAttributeServlet")
public class SetAttributeServlet extends HttpServlet {
private static final long serialVersionUID = 1L;

public void doGet(HttpServletRequest request, HttpServletResponse response) throws ServletException,
IOException {
        doPost(request, response);
    }

public void doPost(HttpServletRequest request, HttpServletResponse response) throws ServletException,
IOException {
        request.setCharacterEncoding("UTF-8");
        response.setContentType("text/html;charset=UTF-8");
        // 获取 Session 对象
        HttpSession session = request.getSession();
        // 构造 HashMap 容器
        HashMap map = new HashMap();
        // 把对象添加到 HashMap 容器里
        map.put("name", "张三丰");
        map.put("address", "湖北省武当山");
        map.put("age", "120");
        // 构造 Vector 容器
        Vector vector = new Vector();
        // 把对象添加到 Vector 容器里
        vector.add("红色");
        vector.add("绿色");
        vector.add("蓝色");
        // 把两个容器对象保存到 Session 对象里
        session.setAttribute("control.map", map);
        session.setAttribute("control.vector", vector);
        PrintWriter out = response.getWriter();
        out.println("<HTML> <HEAD>");
        out.println("<TITLE>给 Servlet 上下文添加属性</TITLE></HEAD>");
        out.println("<BODY>");
        out.println("<a href=GetAttributeServlet>");
        out.println("去到显示 Servlet");
        out.println("<a></BODY>");
        out.println("</HTML>");
        out.flush();
        out.close();
```

```
    }

}
```

示例 6-5：

```
package control;

import java.io.IOException;
import java.io.PrintWriter;
import java.util.HashMap;
import java.util.Vector;
import javax.servlet.ServletException;
import javax.servlet.annotation.WebServlet;
import javax.servlet.http.HttpServlet;
import javax.servlet.http.HttpServletRequest;
import javax.servlet.http.HttpServletResponse;
import javax.servlet.http.HttpSession;

@WebServlet("/GetAttributeServlet")
public class GetAttributeServlet extends HttpServlet {
    private static final long serialVersionUID = 1L;

    public void doGet(HttpServletRequest request, HttpServletResponse response) throws ServletException,
IOException {
        doPost(request, response);
    }

    public void doPost(HttpServletRequest request, HttpServletResponse response) throws ServletException,
IOException {
        request.setCharacterEncoding("UTF-8");
        response.setContentType("text/html;charset=UTF-8");
        PrintWriter out = response.getWriter();
        // 获取 Session 对象
        HttpSession session = request.getSession();
        // 从 Session 对象里获取容器对象
        HashMap map = (HashMap) session.getAttribute("control.map");
        Vector vect = (Vector) session.getAttribute("control.vector");
        out.println("<HTML>");
        out.println(" <HEAD><TITLE>显示 ServletContext</TITLE></HEAD>");
        out.println(" <BODY>");
        out.print("从 SetAttributeServlet 得到对象 1，内容是 ");
        out.print("<table border=1><tr><td>名字</td><td>值</td></tr>");
        out.print("<tr><td>名字：</td><td>" + map.get("name") + "</td></tr>");
        out.print("<tr><td>地址：</td><td>" + map.get("address") + "</td></tr>");
        out.print("<tr><td>年龄：</td><td>" + map.get("age") + "</td></tr>");
        out.print("</table> ");
```

```
        out.print("从 SetAttributeServlet 得到对象 2，内容是 ");
        out.print("<table border=1><tr><td>序号</td><td>值</td></tr>");
        for (int i = 0; i < vect.size(); i++) {
            out.print("<tr><td>" + (i + 1) + "</td><td>" + vect.get(i) + "</td></tr>");
        }
        out.println("</table>");
        out.println(" </BODY>");
        out.println("</HTML>");
        out.flush();
        out.close();
    }
}
```

在 IE 地址栏输入 http://localhost:8080/MySixthWebApp/SetAttributeServlet，显示结果如图 6-1 所示。

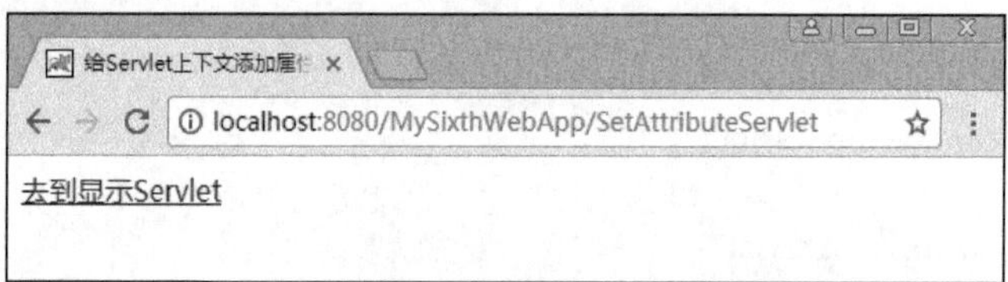

图 6-1

单击“去到显示 Servlet”超链接，显示结果如图 6-2 所示。

图 6-2

这里必须注意的是，使用 Session 来共享数据，必须保证这些页面在同一个会话的生存期内，如果过了这个时间，原来的会话对象会被容器销毁，存放在会话对象中的对象也就不存在了，其他的页面自然从会话对象拿到的都是空对象。其实这种方式在网站购物中使用得最多，用户登录到某个购物站点，便会有个购物车，我们在不同的页面浏览不同的商品，可以把感兴趣的商品全部放入购物车，同时，也可以随时查看已经放入购物车的商品，用户完成一次购物，可以去收银台结账，等等，这些实际上都是通过会话对象来实现的。

6.2　ServletConfig 接口

上述章节讲解了使用 Session 来实现页面之间的通信，我们知道，Session 默认的生存时间是 1800 秒，如果用户在这个时间内没有与服务器交互的动作，那么过了这个时间，Session 就会被服务器销毁。这时我们就会问，有没有生存时间更长的对象来实现页面间或不同的 Servlet 间的通信呢？其实，在 Servlet API 中提供了 ServletContext 接口来实现上述功能，而 ServletConfig 用于获取 Servlet 初始化参数和 ServletContext 对象，因此我们先来学习 ServletConfig 接口。

容器在初始化一个 Servlet 时，将为该 Servlet 创建一个唯一的 ServletConfig 对象，并将初始化参数封装到 ServletConfig 对象中，再调用 Servlet 的 init()方法，将 ServletConfig 对象传递给 Servlet。下面就来学习 Servlet 初始化参数的配置和初始化参数的获取。

6.2.1　Servlet 初始化参数的配置

在实际应用开发中，我们会遇到一些需求不断变更而要改动代码的状况，例如数据库的用户名、密码改变了，若将这些信息编码到 Servlet 类中，则信息的每次变更都将使 Servlet 重新编译，这会大大降低系统的可维护性。这种情况可以采用 Servlet 的初始化参数配置来解决。

下面示例就是通过在 web.xml 中的<servlet>标签下，通过<init-param>标签为 Servlet 配置 mysql 数据库连接的几个初始化参数。代码如示例 6-6 所示。

示例 6-6：

```
<servlet>
    <servlet-name>ServletConfigTest</servlet-name>
    <servlet-class>control.ServletConfigTest</servlet-class>
    <init-param>
        <param-name>url</param-name>
        <param-value>jdbc:mysql://localhost:3306/shoppingcart</param-value>
    </init-param>
    <init-param>
        <param-name>user</param-name>
        <param-value>root</param-value>
    </init-param>
    <init-param>
        <param-name>password</param-name>
        <param-value>root</param-value>
    </init-param>
</servlet>
<servlet-mapping>
<servlet-name>ServletConfigTest</servlet-name>
<url-pattern>/ServletConfigTest</url-pattern>
```

```
</servlet-mapping>
```

在上述示例代码中，<init-param>元素设定初始化参数信息，该元素有两个子元素：<param-name>子元素设置初始化参数名，<param-value>子元素设置初始化参数值。

6.2.2 Servlet 初始化参数的获取

在配置好一些初始化参数后，在应用中如何获取这些参数值？实现代码如示例 6-7 所示。

示例 6-7：

```
package control;

import java.io.IOException;
import java.util.Enumeration;

import javax.servlet.ServletConfig;
import javax.servlet.ServletException;
import javax.servlet.http.HttpServlet;
import javax.servlet.http.HttpServletRequest;
import javax.servlet.http.HttpServletResponse;

public class ServletConfigTest extends HttpServlet {
  public void doGet(HttpServletRequest request, HttpServletResponse response) throws ServletException,
      IOException {

      ServletConfig config = this.getServletConfig(); //  拿到 init()方法中的 ServletConfig 对象

      // --获取当前 Servlet  在 web.xml 中配置的名称(用得不多)
      String sName = config.getServletName();
      System.out.println("当前 Servlet  在 web.xml 中配置的名称:" + sName);

      // --获取当前 Servlet 中配置的初始化参数(只能获取一个)经常用到
      // String url = config.getInitParameter("url");
      // System.out.println(url);

      // --获取当前 Servlet 中配置的初始化参数(全部获取)经常用到
      Enumeration enumration = config.getInitParameterNames();
      while (enumration.hasMoreElements()) {
            String name = (String) enumration.nextElement();
            String value = config.getInitParameter(name);
            System.out.println(name + ":" + value);
      }
  }
```

```
	public void doPost(HttpServletRequest request, HttpServletResponse response) throws ServletException,
IOException {
		doGet(request, response);
	}

}
```

在浏览器输入地址 http://localhost:8080/MySixthWebApp/ServletConfigTest，查看控制台，可以看到控制台打印内容如图 6-3 所示。

```
Markers  Properties  Servers  Data Source Explorer  Snippets  Console  Maven Repositories  Terminal
Tomcat v8.5 Server at localhost [Apache Tomcat] D:\Program Files\Java\jdk1.8.0_161\bin\javaw.exe (2018年8月29日 上午11:36:21)
信息: Deployment of web application directory [D:\apache-tomcat-8.5.32\webapps\manager] has finished in [1,173] ms
八月 29, 2018 11:36:51 上午 org.apache.catalina.startup.HostConfig deployDirectory
信息: Deploying web application directory [D:\apache-tomcat-8.5.32\webapps\ROOT]
八月 29, 2018 11:36:52 上午 org.apache.jasper.servlet.TldScanner scanJars
信息: At least one JAR was scanned for TLDs yet contained no TLDs. Enable debug logging for this logger for a complete
八月 29, 2018 11:36:52 上午 org.apache.catalina.startup.HostConfig deployDirectory
信息: Deployment of web application directory [D:\apache-tomcat-8.5.32\webapps\ROOT] has finished in [1,178] ms
八月 29, 2018 11:36:52 上午 org.apache.coyote.AbstractProtocol start
信息: Starting ProtocolHandler ["http-nio-8080"]
八月 29, 2018 11:36:52 上午 org.apache.coyote.AbstractProtocol start
信息: Starting ProtocolHandler ["ajp-nio-8009"]
八月 29, 2018 11:36:52 上午 org.apache.catalina.startup.Catalina start
信息: Server startup in 29651 ms
当前Servlet 在web.xml中配置的名称:ServletConfigTest
password:root
user:root
url:jdbc:mysql://localhost:3306/shoppingcart
```

图 6-3

6.3　ServletContext 接口

ServletContext 也称为 Servlet 上下文，代表当前的 Servlet 运行环境，是一个全局的储存信息空间，Servlet 容器在启动一个 Web 应用时，会为该应用创建一个唯一的 ServletContext 对象，每个 Servlet 都可以对它进行访问。

6.3.1　ServletContext 获取 Web 项目信息

因为一个 Web 项目只有一个 ServletContext 对象，所以这个对象对整个项目的相关内容都是可以获取的。下面我们学习几个常用的方法，以获取项目的一些信息。

1. 获取文件的 MIME 类型

我们在做文件下载的时候，必须设置两个响应头：Content-Type 和 Content-Disposition。其中 Content-Type 代表的就是文件名称的 MIME 类型，例如常见的 html 文件的 MIME 类型就是“text/html”，JPEG 图片文件的 MIME 类型就是“image/jpeg”，普通文本文档的 MIME 类型就是“text/plain”。我们在 MySixthWebApp 项目下新建一个 Servlet 名为 ServletContextTest，示例 6-8 代码就是用于获取文件的 MIME 类型的主要代码。

示例 6-8：

```
protected void doGet(HttpServletRequest request, HttpServletResponse response)
        throws ServletException, IOException {
    /**
     * 获取文件的 MIME 类型
     *
     * 1. 先获取 servletContext 2.通过 getMimeType(String file);
     */
    ServletContext servletContext = this.getServletContext();
    String mimeType = servletContext.getMimeType("mime.txt");
    System.out.println("mime.txt 的 MIME 类型是：" + mimeType);
}
```

在浏览器输入地址 http://localhost:8080/MySixthWebApp/ServletContextTest，查看控制台打印信息如图 6-4 所示。

```
Markers  Properties  Servers  Data Source Explorer  Snippets  Console  Maven Repositories  Terminal
Tomcat v8.5 Server at localhost [Apache Tomcat] D:\Program Files\Java\jdk1.8.0_161\bin\javaw.exe (2018年8月29日 下午2:41:48)
八月 29, 2018 2:42:19 下午 org.apache.jasper.servlet.TldScanner scanJars
信息: At least one JAR was scanned for TLDs yet contained no TLDs. Enable debug logging for this logger for a complete
八月 29, 2018 2:42:19 下午 org.apache.catalina.startup.HostConfig deployDirectory
信息: Deployment of web application directory [D:\apache-tomcat-8.5.32\webapps\manager] has finished in [1,272] ms
八月 29, 2018 2:42:19 下午 org.apache.catalina.startup.HostConfig deployDirectory
信息: Deploying web application directory [D:\apache-tomcat-8.5.32\webapps\ROOT]
八月 29, 2018 2:42:20 下午 org.apache.jasper.servlet.TldScanner scanJars
信息: At least one JAR was scanned for TLDs yet contained no TLDs. Enable debug logging for this logger for a complete
八月 29, 2018 2:42:20 下午 org.apache.catalina.startup.HostConfig deployDirectory
信息: Deployment of web application directory [D:\apache-tomcat-8.5.32\webapps\ROOT] has finished in [1,233] ms
八月 29, 2018 2:42:20 下午 org.apache.coyote.AbstractProtocol start
信息: Starting ProtocolHandler ["http-nio-8080"]
八月 29, 2018 2:42:20 下午 org.apache.coyote.AbstractProtocol start
信息: Starting ProtocolHandler ["ajp-nio-8009"]
八月 29, 2018 2:42:20 下午 org.apache.catalina.startup.Catalina start
信息: Server startup in 29763 ms
mime.txt的MIME类型是: text/plain
```

图 6-4

2. 获取 web 项目请求工程名

ServletContext 的 getContextPath()方法是用于返回当前 Web 应用的根路径，如示例 6-9 所示，运行结果如图 6-5 所示。

示例 6-9：

```
/**
 * 获取文件的项目的根路径
 *
 * 1. 先获取 servletContext 2.通过 getContextPath()方法获取根路径
 */
ServletContext servletContext = this.getServletContext();


String path = servletContext.getContextPath();
System.out.println("项目路径名：" + path);
```

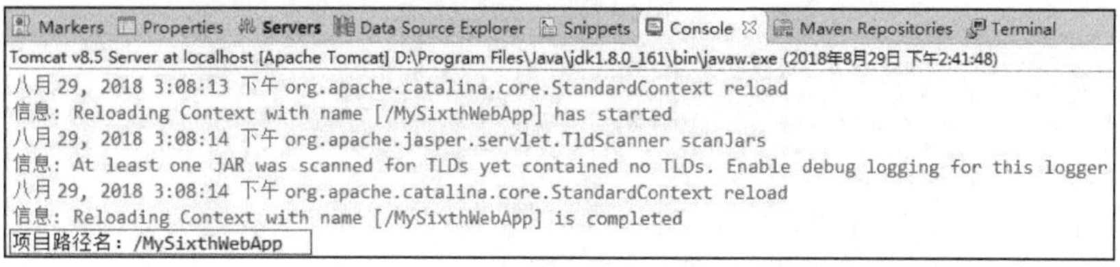

```
Markers  Properties  Servers  Data Source Explorer  Snippets  Console ⊠  Maven Repositories  Terminal
Tomcat v8.5 Server at localhost [Apache Tomcat] D:\Program Files\Java\jdk1.8.0_161\bin\javaw.exe (2018年8月29日 下午2:41:48)
八月 29, 2018 3:08:13 下午 org.apache.catalina.core.StandardContext reload
信息: Reloading Context with name [/MySixthWebApp] has started
八月 29, 2018 3:08:14 下午 org.apache.jasper.servlet.TldScanner scanJars
信息: At least one JAR was scanned for TLDs yet contained no TLDs. Enable debug logging for this logger
八月 29, 2018 3:08:14 下午 org.apache.catalina.core.StandardContext reload
信息: Reloading Context with name [/MySixthWebApp] is completed
项目路径名：/MySixthWebApp
```

图 6-5

3. 获取 Web 项目的初始化参数

首先在 web.xml 配置文件中配置 Web 应用范围的初始化参数，该参数通过<content-param>元素来指定，代码如示例 6-10 所示。

示例 6-10：

```
<context-param>
    <param-name>webSite</param-name>
    <param-value>http://www.myhopu.com/</param-value>
</context-param>
<context-param>
    <param-name>email</param-name>
    <param-value>hopu@myhopu.com</param-value>
</context-param>
```

然后通过 ServletContext 对象获取初始化参数的值，主要代码如示例 6-11 所示。控制台运行结果如图 6-6 所示。

示例 6-11：

```
ServletContext servletContext = this.getServletContext();
    // 使用 ServletContext 对象获取某个参数的值
    String webSite = servletContext.getInitParameter("webSite");
    String email = servletContext.getInitParameter("email");
    System.out.println("webSite 参数的值：" + webSite);
    System.out.println("email 参数的值：" + email);
    System.out.println("-----------------------------");
    // 使用 ServletContext 对象获取所有初始化参数
    Enumeration<String> parameterNames = servletContext.getInitParameterNames();
    while (parameterNames.hasMoreElements()) {
        String name = parameterNames.nextElement();
        String value = servletContext.getInitParameter(name);
        System.out.println(name + ":" + value);
    }
```

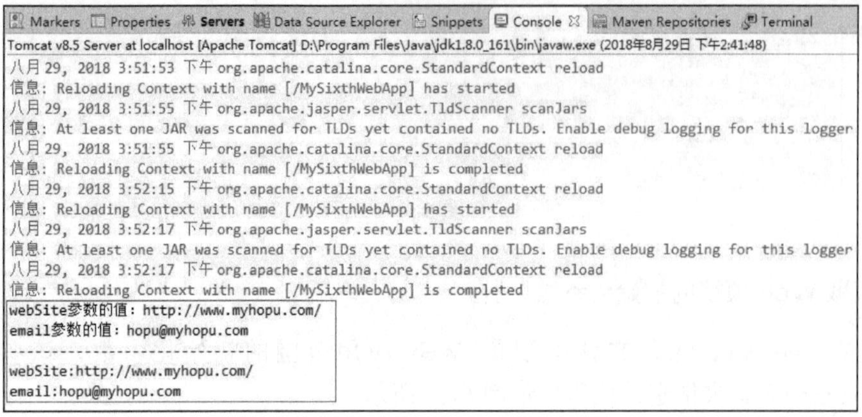

```
Markers  Properties  Servers  Data Source Explorer  Snippets  Console ⊠  Maven Repositories  Terminal
Tomcat v8.5 Server at localhost [Apache Tomcat] D:\Program Files\Java\jdk1.8.0_161\bin\javaw.exe (2018年8月29日 下午2:41:48)
八月 29, 2018 3:51:53 下午 org.apache.catalina.core.StandardContext reload
信息: Reloading Context with name [/MySixthWebApp] has started
八月 29, 2018 3:51:55 下午 org.apache.jasper.servlet.TldScanner scanJars
信息: At least one JAR was scanned for TLDs yet contained no TLDs. Enable debug logging for this logger
八月 29, 2018 3:51:55 下午 org.apache.catalina.core.StandardContext reload
信息: Reloading Context with name [/MySixthWebApp] is completed
八月 29, 2018 3:52:15 下午 org.apache.catalina.core.StandardContext reload
信息: Reloading Context with name [/MySixthWebApp] has started
八月 29, 2018 3:52:17 下午 org.apache.jasper.servlet.TldScanner scanJars
信息: At least one JAR was scanned for TLDs yet contained no TLDs. Enable debug logging for this logger
八月 29, 2018 3:52:17 下午 org.apache.catalina.core.StandardContext reload
信息: Reloading Context with name [/MySixthWebApp] is completed
webSite参数的值: http://www.myhopu.com/
email参数的值: hopu@myhopu.com
------------------------------
webSite:http://www.myhopu.com/
email:hopu@myhopu.com
```

图 6-6

6.3.2 ServletContext 读取服务器文件资源

使用 ServletContext 接口可以直接访问 Web 应用中的静态内容文件，也可以读取文件在服务器中的真实存放路径。示例 6-12 演示了读取数据库文件 db.properties 文件。控制台运行结果如图 6-7 所示。

示例 6-12：

```
ServletContext servletContext = this.getServletContext();
    // getRealPath(String path),获取文件在系统磁盘上的真实路径
    String path = servletContext.getRealPath("db.properties");
    System.out.println("path:" + path);
    // getResourceAsStream(String path)，获取指定文件的输入流，参数需要以"/"开头
    // 需要注意的是，项目的目录结构和部署在服务器上的目录结构并不一致，src 对应的是
/WEB-INF/classes
        InputStream inputStream =
servletContext.getResourceAsStream("/WEB-INF/classes/db.properties");
        Properties properties = new Properties();
        // 从输入流中读取属性列表
        properties.load(inputStream);
        String url = properties.getProperty("url");
        String user = properties.getProperty("username");
        String password = properties.getProperty("password");
        System.out.println("url:" + url);
        System.out.println("username:" + user);
        System.out.println("password:" + password);
```

```
Markers  Properties  Servers  Data Source Explorer  Snippets  Console  Maven Repositories  Terminal
Tomcat v8.5 Server at localhost [Apache Tomcat] D:\Program Files\Java\jdk1.8.0_161\bin\javaw.exe (2018年8月29日 下午5:14:52)
信息: Deployment of web application directory [D:\apache-tomcat-8.5.32\webapps\manager] has finished in [1,453] ms
八月 29, 2018 5:15:27 下午 org.apache.catalina.startup.HostConfig deployDirectory
信息: Deploying web application directory [D:\apache-tomcat-8.5.32\webapps\ROOT]
八月 29, 2018 5:15:28 下午 org.apache.jasper.servlet.TldScanner scanJars
信息: At least one JAR was scanned for TLDs yet contained no TLDs. Enable debug logging for this logger for a complete
八月 29, 2018 5:15:28 下午 org.apache.catalina.startup.HostConfig deployDirectory
信息: Deployment of web application directory [D:\apache-tomcat-8.5.32\webapps\ROOT] has finished in [1,164] ms
八月 29, 2018 5:15:28 下午 org.apache.coyote.AbstractProtocol start
信息: Starting ProtocolHandler ["http-nio-8080"]
八月 29, 2018 5:15:28 下午 org.apache.coyote.AbstractProtocol start
信息: Starting ProtocolHandler ["ajp-nio-8009"]
八月 29, 2018 5:15:28 下午 org.apache.catalina.startup.Catalina start
信息: Server startup in 31341 ms
path:F:\svse\MySixthWebApp\WebContent\db.properties
url:jdbc:mysql://localhost:3306/shoppingcart?useUnicode=true&characterEncoding=utf-8
username:root
password:123456
```

图 6-7

6.3.3　ServletContext 作为域对象存取数据

域对象指的是将数据存入到域对象中，这个数据就会有一定的作用范围。域指的是一定的作用范围，作用是保存、获取数据，可以在不同的动态资源之间共享数据。具体方法如表 6-1 所示。

表 6-1

方法	方法描述
setAttribute(String name,Object object)	向域对象中存入数据的方法，没有返回值
getAttribute(String name)	根据 name 获取对应的数据，返回的是 object 对象
removeAttribute(String name)	根据指定 name 移除数据的方法

下面代码通过创建两个 Servlet 来演示域对象的存取方法。如示例 6-13 和示例 6-14 所示。

示例 6-13：

```
package control;

import java.io.IOException;

import javax.servlet.ServletException;
import javax.servlet.annotation.WebServlet;
import javax.servlet.http.HttpServlet;
import javax.servlet.http.HttpServletRequest;
import javax.servlet.http.HttpServletResponse;

@WebServlet("/ContextAttributeServlet")
public class ContextAttributeServlet extends HttpServlet {
    private static final long serialVersionUID = 1L;

    @Override
    public void init() throws ServletException {
        // 给当前 Servlet 初始化一个值
```

```
        this.getServletContext().setAttribute("name", "厚溥");
    }

    protected void doGet(HttpServletRequest request, HttpServletResponse response)
            throws ServletException, IOException {
        // 访问该 Servlet 时取出存入的值，打印在控制台
        String name = (String) this.getServletContext().getAttribute("name");
        System.out.println("访问第一 Servlet 控制台打印姓名：" + name);
    }

    protected void doPost(HttpServletRequest request, HttpServletResponse response)
            throws ServletException, IOException {
        doGet(request, response);
    }

}
```

示例 6-14：

```
package control;

import java.io.IOException;

import javax.servlet.ServletException;
import javax.servlet.annotation.WebServlet;
import javax.servlet.http.HttpServlet;
import javax.servlet.http.HttpServletRequest;
import javax.servlet.http.HttpServletResponse;

@WebServlet("/ContextAttributeOtherServlet")
public class ContextAttributeOtherServlet extends HttpServlet {
    private static final long serialVersionUID = 1L;

    protected void doGet(HttpServletRequest request, HttpServletResponse response)
            throws ServletException, IOException {
        // 访问当前 Servlet 时，同样取出存入的值，查看控制台是否有打印
        String name = (String) this.getServletContext().getAttribute("name");
        System.out.println("访问第二 Servlet 控制台打印姓名：" + name);
    }

    protected void doPost(HttpServletRequest request, HttpServletResponse response)
            throws ServletException, IOException {
        doGet(request, response);
    }

}
```

我们先启动服务器在浏览器访问 http://localhost:8080/MySixthWebApp/ContextAttributeServlet，再访问 http://localhost:8080/MySixthWebApp/ContextAttributeOtherServlet。前后运行结果如图 6-8 和图 6-9 所示。

```
Markers  Properties  Servers  Data Source Explorer  Snippets  Console  Maven Repositories  Terminal
Tomcat v8.5 Server at localhost [Apache Tomcat] D:\Program Files\Java\jdk1.8.0_161\bin\javaw.exe (2018年8月31日 下午2:47:04)
信息: Deploying web application directory [D:\apache-tomcat-8.5.32\webapps\manager]
八月 31, 2018 2:47:37 下午 org.apache.jasper.servlet.TldScanner scanJars
信息: At least one JAR was scanned for TLDs yet contained no TLDs. Enable debug logging for this logger for a complete
八月 31, 2018 2:47:37 下午 org.apache.catalina.startup.HostConfig deployDirectory
信息: Deployment of web application directory [D:\apache-tomcat-8.5.32\webapps\manager] has finished in [1,251] ms
八月 31, 2018 2:47:37 下午 org.apache.catalina.startup.HostConfig deployDirectory
信息: Deploying web application directory [D:\apache-tomcat-8.5.32\webapps\ROOT]
八月 31, 2018 2:47:38 下午 org.apache.jasper.servlet.TldScanner scanJars
信息: At least one JAR was scanned for TLDs yet contained no TLDs. Enable debug logging for this logger for a complete
八月 31, 2018 2:47:38 下午 org.apache.catalina.startup.HostConfig deployDirectory
信息: Deployment of web application directory [D:\apache-tomcat-8.5.32\webapps\ROOT] has finished in [1,248] ms
八月 31, 2018 2:47:38 下午 org.apache.coyote.AbstractProtocol start
信息: Starting ProtocolHandler ["http-nio-8080"]
八月 31, 2018 2:47:38 下午 org.apache.coyote.AbstractProtocol start
信息: Starting ProtocolHandler ["ajp-nio-8009"]
八月 31, 2018 2:47:38 下午 org.apache.catalina.startup.Catalina start
信息: Server startup in 31219 ms
访问第一Servlet控制台打印姓名：厚溥
```

图 6-8

```
Markers  Properties  Servers  Data Source Explorer  Snippets  Console  Maven Repositories  Terminal
Tomcat v8.5 Server at localhost [Apache Tomcat] D:\Program Files\Java\jdk1.8.0_161\bin\javaw.exe (2018年8月31日 下午2:47:04)
八月 31, 2018 2:47:37 下午 org.apache.jasper.servlet.TldScanner scanJars
信息: At least one JAR was scanned for TLDs yet contained no TLDs. Enable debug logging for this logger for a complete
八月 31, 2018 2:47:37 下午 org.apache.catalina.startup.HostConfig deployDirectory
信息: Deployment of web application directory [D:\apache-tomcat-8.5.32\webapps\manager] has finished in [1,251] ms
八月 31, 2018 2:47:37 下午 org.apache.catalina.startup.HostConfig deployDirectory
信息: Deploying web application directory [D:\apache-tomcat-8.5.32\webapps\ROOT]
八月 31, 2018 2:47:38 下午 org.apache.jasper.servlet.TldScanner scanJars
信息: At least one JAR was scanned for TLDs yet contained no TLDs. Enable debug logging for this logger for a complete
八月 31, 2018 2:47:38 下午 org.apache.catalina.startup.HostConfig deployDirectory
信息: Deployment of web application directory [D:\apache-tomcat-8.5.32\webapps\ROOT] has finished in [1,248] ms
八月 31, 2018 2:47:38 下午 org.apache.coyote.AbstractProtocol start
信息: Starting ProtocolHandler ["http-nio-8080"]
八月 31, 2018 2:47:38 下午 org.apache.coyote.AbstractProtocol start
信息: Starting ProtocolHandler ["ajp-nio-8009"]
八月 31, 2018 2:47:38 下午 org.apache.catalina.startup.Catalina start
信息: Server startup in 31219 ms
访问第一Servlet控制台打印姓名：厚溥
访问第二Servlet控制台打印姓名：厚溥
```

图 6-9

6.4 HttpServletRequest 接口

在 Servlet API 中，ServletRequest 接口被定义为用于封装请求的信息，ServletRequest 对象由 Servlet 容器在用户每次请求 Servlet 时创建并传入 Servlet 的 service()方法中。通过这个对象提供的方法，可以获得客户端请求的所有信息。

6.4.1 获取请求行信息

一个 HTTP 请求报文由请求行(request line)、请求头部(header)、空行和请求数据 4 个部分组成，图 6-10 给出了请求报文的一般格式。请求行分为 3 个部分：请求方法、请求地址和协议版本。

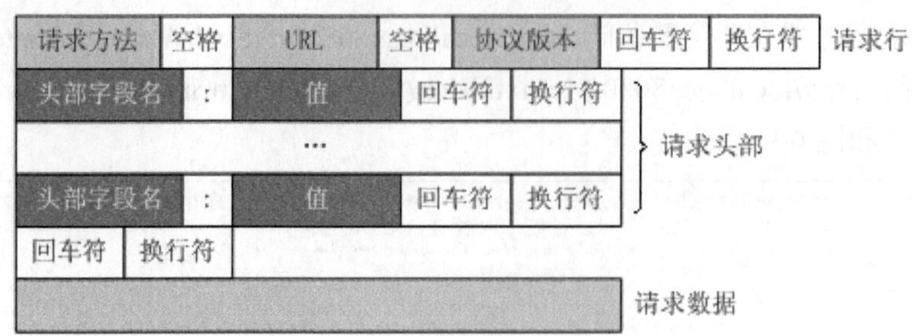

图 6-10

HttpServletRequest 接口对请求行各部分信息的获取方法如表 6-2 所示。

表 6-2

方法	描述
getMethod()	获取请求使用的 HTTP 方法，如 GET、POST 和 PUT
getRequestURL()	获取完整请求路径
getRequestURI()	获取请求行中除了域名外的资源名部分
getProtocol()	获取使用的协议及版本号
getQueryString()	获取请求 URL 后面的查询字符串，只对 GET 有效
getContextPath()	获取请求 Web 应用名称

下面我们用代码演示 HttpServletRequest 接口获取请求行信息的方法的使用。如示例 6-15 所示。

示例 6-15：

```
package control;

import java.io.IOException;
import java.io.PrintWriter;

import javax.servlet.ServletException;
import javax.servlet.annotation.WebServlet;
import javax.servlet.http.HttpServlet;
import javax.servlet.http.HttpServletRequest;
import javax.servlet.http.HttpServletResponse;

@WebServlet("/RequestLineServlet")
public class RequestLineServlet extends HttpServlet {
    private static final long serialVersionUID = 1L;

    protected void doGet(HttpServletRequest request, HttpServletResponse response)
            throws ServletException, IOException {
        /**
         * 1.获得客户机信息
```

```
    */
    String requestUrl = request.getRequestURL().toString();// 得到请求的 URL 地址
    String requestUri = request.getRequestURI();// 得到请求的资源
    String queryString = request.getQueryString();// 得到请求的 URL 地址中附带的参数
    String method = request.getMethod();// 得到请求 URL 地址时使用的方法
    String protocol = request.getProtocol();// 获取使用的协议和版本号
    String contextPath = request.getContextPath();// 获取请求资源 Web 应用的路径
    response.setContentType("text/html;charset=utf-8");// 设置将字符以"UTF-8"编码输出到客户端浏
                                                         览器
    PrintWriter out = response.getWriter();
    out.println("<p>获取到的客户机信息如下：</p>");
    out.println("<hr/>");
    out.println("<p>请求的 URL 地址：" + requestUrl + "</p>");
    out.println("<p>请求的资源：" + requestUri + "</p>");
    out.println("<p>请求的 URL 地址中附带的参数：" + queryString + "</p>");
    out.println("<p>请求使用的方法：" + method + "</p>");
    out.println("<p>请求使用的协议和版本号：" + protocol + "</p>");
    out.println("<p>请求资源 Web 应用的路径:" + contextPath + "</p>");

  }

  protected void doPost(HttpServletRequest request, HttpServletResponse response)
          throws ServletException, IOException {
    doGet(request, response);
  }

}
```

启动服务器，在浏览器中输入网址 http://localhost:8080/MySixthWebApp/RequestLineServlet?-pageNo=2&queryString=hopu，查看页面结果，如图 6-11 所示。

```
localhost:8080/MySixth ×
localhost:8080/MySixthWebApp/RequestLineServlet?pageNo=2&queryString=hopu
获取到的客户机信息如下：

请求的URL地址：http://localhost:8080/MySixthWebApp/RequestLineServlet
请求的资源：/MySixthWebApp/RequestLineServlet
请求的URL地址中附带的参数：pageNo=2&queryString=hopu
请求使用的方法：GET
请求使用的协议和版本号：HTTP/1.1
请求资源web应用的路径:/MySixthWebApp
```

图 6-11

6.4.2　获取请求头信息

我们在前面通过浏览器访问 url 时，通过检查或者按 F12 快捷键，可以查看到页面的请求信息，其中就包含请求头，如图 6-12 所示。

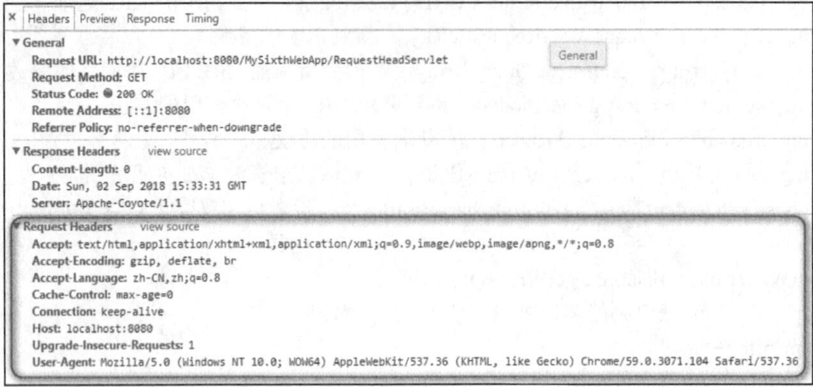

图 6-12

在 Servlet 中，可以通过 HttpServletRequest 的 getHeader(string name)方法和 String getHeaders(String name)方法或者 Enumeration getHeaderNames()方法来获取客户端请求头信息，代码如示例 6-16 所示。

示例 6-16：

```
package control;

import java.io.IOException;
import java.util.Enumeration;

import javax.servlet.ServletException;
import javax.servlet.annotation.WebServlet;
import javax.servlet.http.HttpServlet;
import javax.servlet.http.HttpServletRequest;
import javax.servlet.http.HttpServletResponse;

@WebServlet("/RequestHeadServlet")
public class RequestHeadServlet extends HttpServlet {
    private static final long serialVersionUID = 1L;

        public RequestHeadServlet() {
            super();
        }

    protected void doGet(HttpServletRequest request, HttpServletResponse response) throws
ServletException, IOException {
        //获取请求头
        String header = request.getHeader("user-Agent");
        System.out.println("请求头："+header);
            System.out.println("------------------------------------------------------");
    //获取所有的消息头名称
```

```
        Enumeration<String> headerNames = request.getHeaderNames();
        //获取的消息头名称，获取对应的值，并输出
        while(headerNames.hasMoreElements()){
            String nextElement = headerNames.nextElement();
            System.out.println(nextElement+":"+request.getHeader(nextElement));
        }
        System.out.println("--------------------------------------------------------");
        //根据名称获取此重名的所有数据
        Enumeration<String> headers = request.getHeaders("accept");
        while (headers.hasMoreElements()) {
            String string = (String) headers.nextElement();
            System.out.println(string);
        }
    }

    protected void doPost(HttpServletRequest request, HttpServletResponse response) throws
ServletException, IOException {
        doGet(request, response);
    }

}
```

启动服务器，在浏览器中输入 http://localhost:8080/MySixthWebApp/RequestHeadServlet 地址，运行结果如图 6-13 所示。

```
Markers  Properties  Servers  Data Source Explorer  Snippets  Console  Progress  Maven Repositories
Tomcat v8.5 Server at localhost [Apache Tomcat] D:\Program Files\Java\jre1.8.0_181\bin\javaw.exe (2018年9月3日 上午12:00:20)
请求头: Mozilla/5.0 (Windows NT 10.0; WOW64) AppleWebKit/537.36 (KHTML, like Gecko) Chrome/59.0.3071.104 Safari/537.36
---------------------------------------------------------
host:localhost:8080
connection:keep-alive
cache-control:max-age=0
upgrade-insecure-requests:1
user-agent:Mozilla/5.0 (Windows NT 10.0; WOW64) AppleWebKit/537.36 (KHTML, like Gecko) Chrome/59.0.3071.104 Safari/537.36
accept:text/html,application/xhtml+xml,application/xml;q=0.9,image/webp,image/apng,*/*;q=0.8
accept-encoding:gzip, deflate, br
accept-language:zh-CN,zh;q=0.8
---------------------------------------------------------
text/html,application/xhtml+xml,application/xml;q=0.9,image/webp,image/apng,*/*;q=0.8
```

图 6-13

6.4.3　作为域对象存取数据

HttpServletRequest 对象代表客户端的请求，当客户端通过 HTTP 协议访问服务器时，HTTP 请求头中的所有信息都封装在这个对象中，通过这个对象提供的方法，可以获得客户端请求的所有信息。

HttpServletRequest 接口存取数据的方法如表 6-3 所示，大家只要掌握了如何使用这些方法学习起来就很容易，在此就不再过多赘述。

表 6-3

方法	描述
void setAttribute(String name,Object value)	向 request 域中存入值
Object getAttribute(String name)	从 request 域中获取 name 属性的值
void removeAttribute(String name)	从 request 域中移除 name 属性的值

6.5 HttpServletResponse 接口

HttpServletResponse 对象代表服务器的响应。这个对象中封装了向客户端发送数据、响应头、响应状态码的方法。查看 HttpServletResponse 的 API，可以看到这些相关的方法。下面将依次对其进行介绍。

6.5.1 关于响应状态

一个完整的 HTTP 响应报文由响应行、响应头、响应体组成，如图 6-14 所示为一个响应报文信息格式样例。

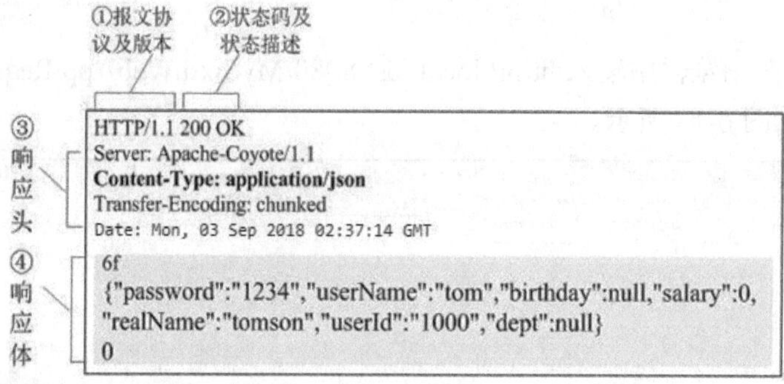

图 6-14

HTTP 协议响应报文的响应行由报文协议及版本和状态码及状态描述构成。状态码由三个十进制数字组成，第一个十进制数字定义了状态码的类型，后面两个数字没有分类的作用。HTTP 状态码共分为 5 种，如表 6-4 所示。

表 6-4

分类	分类描述
1**	表示接收到请求并且继续处理
2**	表示动作被成功接收、理解和接受
3**	为了完成指定的动作，必须接受进一步处理
4**	请求包含错误语法或不能正确执行
5**	服务器不能正确执行一个正确的请求

常见的响应状态码有：

- 200 表明该请求被成功地完成，所请求的资源发送回客户端。
- 302 请求的网页被转移到一个新的地址，但客户访问仍继续通过原始 URL 地址重定向，新的 URL 会在 response 中的 Location 中返回，浏览器将会使用新的 URL 发出新的 Request。
- 404 表示请求资源不存在。
- 500 表示服务器内部错误。

6.5.2 关于响应头方法

在 Servlet 中，可以通过 HttpServletResponse 的 setHeader()方法来设置 HTTP 响应消息头，通过表 6-5 我们来看一下常用的 HTTP 响应消息头。

表 6-5

响应报头名称	说明
Location: http://www.myhopu.com/	表示重定向的地址，该头和 302 的状态码一起使用
Server:apache tomcat	表示服务器的类型
Content-Encoding: gzip	表示服务器发送给浏览器的数据压缩类型
Content-Length: 80	表示服务器发送给浏览器的数据长度
Content-Language: zh-cn	表示服务器支持的语言
Content-Type: text/html; charset=UTF-8	表示服务器发送给浏览器的数据类型及内容编码
Last-Modified: Tue, 11 Jul 2000 18:23:51 GMT	表示服务器资源的最后修改时间
Refresh: 1;	表示定时刷新
Content-Disposition: attachment; filename=aaa.zip	表示告诉浏览器以下载方式打开资源(下载文件时用到)
Transfer-Encoding: chunked	表示为了安全传输而对实体进行的编码
Set-Cookie:SS=Q0=5Lb_nQ; path=/search	表示服务器发送给浏览器的 cookie 信息(会话管理用到)
Expires: -1	表示通知浏览器不再进行缓存
Cache-Control: no-cache	缓存机制，no-cache 阻止浏览器缓存页面
Pragma: no-cache	禁止缓存当前文档内容
Connection: Close/Keep-Alive	表示服务器和浏览器的连接状态。close：关闭连接；keep-alive：保持连接

对于一些常用的消息头，Servlet API 中也提供了一些特定的方法来进行设置，如表 6-6 所示。

表 6-6

响应方法	说明
setContentType(String mime)	设定 Content-Type 消息头
setContentLength(int length)	设定 Content-Length 消息头
addHeader(String name,String value)	新增 String 类型的值到名为 name 的 http 头部

(续表)

响应方法	说明
addIntHeader(String name,int value)	新增 int 类型的值到名为 name 的 http 头部
addDateHeader(String name,long date)	新增 long 类型的值到名为 name 的 http 头部
addCookie(Cookie c)	为 Set-Cookie 消息头增加一个值

6.5.3 关于响应体方法

在 Servlet 中，向客户端输出的响应数据是通过输出流对象来完成的，HttpServletResponse 接口提供了两个获取不同类型输出流对象的方法，一个是 getOutputStream()方法，返回字节输出流 ServletOutputStream 对象；一个是 getWriter()方法，返回字符输出流 PrintWriter 对象。

ServletOutputStream 对象主要用于输出二进制字节流数据，PrintWriter 对象主要用于输出字符文本内容，但其内部实现依旧是将字符串转换成了某种字符集编码的字节数组再进行输出。ServletOutputStream 对象虽然也可以输出文本字符，但是 PrintWriter 对象更易于完成文本到字节数组的转换。

当向 ServletOutputStream 或 PrintWriter 对象中写入数据后，这些数据将被 Servlet 引擎从 response 中获取，Servlet 引擎将这些数据当作响应消息的正文，然后再与响应状态行和各响应头组合后输出到客户端。

示例 6-17 演示了如何使用 ServletOutputStream 对象响应输出一个服务器端的图形。

示例 6-17：

```
package control;

import java.io.IOException;
import java.io.InputStream;

import javax.servlet.ServletContext;
import javax.servlet.ServletException;
import javax.servlet.ServletOutputStream;
import javax.servlet.annotation.WebServlet;
import javax.servlet.http.HttpServlet;
import javax.servlet.http.HttpServletRequest;
import javax.servlet.http.HttpServletResponse;

@WebServlet("/OutputStreamServlet")
public class OutputStreamServlet extends HttpServlet {
  protected void doGet(HttpServletRequest request, HttpServletResponse response)
          throws ServletException, IOException {
     // 设置响应消息头
     response.setContentType("image/jpg");
```

```
        // 获取 ServletContext 对象
        ServletContext context = this.getServletContext();
        // 获取读取服务器端文件的输入流
        InputStream inputStream = context.getResourceAsStream("/WEB-INF/images/hopu.jpg");
        // 获取 ServletOutputStream 输出流
        ServletOutputStream outputStream = response.getOutputStream();
        int i = 0;
        while ((i = inputStream.read()) != -1) {
        // 向输出流中写入二进制数据
            outputStream.write(i);
        }
        inputStream.close();
        outputStream.close();
    }

    protected void doPost(HttpServletRequest request, HttpServletResponse response)
            throws ServletException, IOException {
        doGet(request, response);
    }
}
```

启动服务器，在浏览器访问 http://localhost:8080/MySixthWebApp/OutputStreamServlet，运行结果如图 6-15 所示。

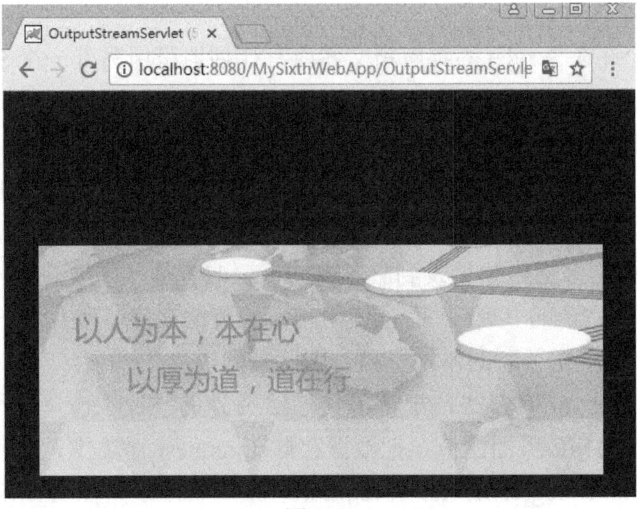

图 6-15

6.6　Servlet 控制器的作用

在实际应用中，Servlet 除了打印页面的内容外，更多的时候是被实现成控制功能，我们习惯把充当控制角色的 Servlet 叫作控制器(Controller)。Servlet API 中定义了把请求转发到同一个 Web 应用中的其他资源、引用其他资源内容或转发到 Web 应用外的标准方法，使 Servlet 被实现成控制器成为可能。它依赖于 RequestDispatcher 对象的 forward()方法和 HttpServletResponse 对象的 sendRedirect()方法，我们称前者是页面间的请求转发，后者是服务器端的请求重定向。下面详细介绍这两个对象的方法。

6.6.1 RequestDispatcher 接口

RequestDispatcher 是 javax.servlet 中定义的接口，实现这个接口的对象允许将请求转发到服务器中的其他资源。RequestDispatcher 接口定义了 forward()和 include()两个方法来转发请求并引用数据。

1. forward()方法

forward()方法允许用户有计划地把请求转发到另外一个代理的 Servlet(或者其他服务器资源)上。转发会把所有控制传递给新资源，并由这个代理来生成响应。要注意的是，必须把请求和响应对象都传递给代理 Servlet，以便代理 Servlet 可以从请求中读取信息并生成适当的响应。我们不仅可以把一个查询字符串放在 URL 中，当把请求传递给代理时，代理可用 request 对象的 getParameter()方法去读取这些信息；而且可以使用 ServletRequest 对象的 setAttribute()方法，把任意多的对象以名-值对的绑定方式存储在 request 对象中，然后传递给代理。代理可以使用 request 对象的 getAttribute()方法取得这些对象。

2. include()方法

include()方法允许在 Servlet 的响应正文体内有计划地引用另外一个 Servlet 所生成的内容，相当于两个 Servlet 合作生成响应。

那么，如何获得 RequestDispatcher 对象的引用呢？有两种方式，第一种是使用 ServletContext 对象的 getRequestDispatcher()方法来获得，第二种是使用 HttpServletRequest 对象的 getRequestDispatcher()方法来获得。两种方式的 getRequestDispatcher()方法的入参是一个 URL 路径，由它来指定要转发到服务器的资源或者是从中要获取内容的资源。URL 路径引用不推荐使用完全的 URL，而是使用相对的资源名字。使用第一种方式时，URL 是与 ServletContext 目录相关联的本地资源，因此必须以“/”开始。

下面我们使用 forward()方法，把请求传递给另外的 Servlet。写两个 Servlet，其中 GetInfoServlet 检索 HTML 页面的值，把数据存储在 request 对象中，然后再把请求转到第二个 DelegateServlet。DelegateServlet 从 request 对象中取得存储在 HTML 表单中的数据，并从数据库中得到更详细的信息，显示在页面上。

我们按下面的步骤来演示上述功能。

(1) 建立一个名叫 input.html 的 HTML 网页，接受用户的输入，代码如示例 6-18 所示。

示例 6-18：

```
<!DOCTYPE html>
<html>
<head>
```

```
<meta charset="UTF-8">
<title>书的详细信息</title>
</head>
<body>
<h1>欢迎来到网上书店</h1> <br>
    请输入书的 ID，以显示详细信息
    <form action="GetInfoServlet" method="POST" >
    <input type="text" name="bookid" value=""><br>
    <input type="submit" value="提交" >
    </form>
</body>
</html>
```

(2) 建立一个名叫 GetInfoServlet 的 Servlet，负责从 input.html 中取得用户输入的 bookid，并把它放到 request 的属性中。然后使用 RequestDispatcher 对象的 forward()方法把控制权转到 DelegateServlet，代码如示例 6-19 所示。

示例 6-19：

```
package control;
import java.io.IOException;
import javax.servlet.RequestDispatcher;
import javax.servlet.ServletContext;
import javax.servlet.ServletException;
import javax.servlet.annotation.WebServlet;
import javax.servlet.http.HttpServlet;
import javax.servlet.http.HttpServletRequest;
import javax.servlet.http.HttpServletResponse;
@WebServlet("/GetInfoServlet")
public class GetInfoServlet extends HttpServlet {
  public void doGet(HttpServletRequest request, HttpServletResponse response) throws ServletException,
IOException {
      doPost(request, response);
  }

  public void doPost(HttpServletRequest request, HttpServletResponse response) throws ServletException,
IOException {
      request.setCharacterEncoding("UTF-8");
      response.setContentType("text/html;charset=UTF-8");
      // 把 bookid 存放在 request 对象中
      String bookid = request.getParameter("bookid");
      request.setAttribute("control.bookid", bookid);
      // 使用 ServletContext 获取 RequestDispatcher 的引用
      // 1.获取 Servlet 上下文
      ServletContext ctx = this.getServletContext();
      // 2.获取 RequestDispatcher 对象，注意 URL 要带有/
      RequestDispatcher rd = null;
```

```
        rd = ctx.getRequestDispatcher("/DelegateServlet?name=bush");
        /*
         * 使用 request 对象去获取 RequestDispatcher 的引用，URL 中可以使用/，也可以不用，效果
         是一样的 RequestDispatcher
         * rd = null; rd = request.getRequestDispatcher("/delegateServlet?name= bush");
         */
        // 转移控制权
        rd.forward(request, response);
    }
}
```

(3) 建立名叫 DelegateServlet 的 Servlet，负责显示信息，代码如示例 6-20 所示。

示例 6-20：

```
package control;

import java.io.IOException;
import java.io.PrintWriter;
import javax.servlet.ServletException;
import javax.servlet.annotation.WebServlet;
import javax.servlet.http.HttpServlet;
import javax.servlet.http.HttpServletRequest;
import javax.servlet.http.HttpServletResponse;

@WebServlet("/DelegateServlet")
public class DelegateServlet extends HttpServlet {
    public void doPost(HttpServletRequest request, HttpServletResponse response) throws ServletException,
    IOException {
        request.setCharacterEncoding("UTF-8");
        response.setContentType("text/html;charset=UTF-8");
        String bookid = (String) request.getAttribute("control.bookid");
        String author = request.getParameter("name");
        PrintWriter out = response.getWriter();
        out.println("<HTML>");
        out.println(" <HEAD><TITLE>显示信息</TITLE></HEAD>");
        out.println(" <BODY>");
        out.print("你输入的书的 ISBN 号码是: " + bookid + "<br>");
        out.print("作者是: " + author + "<br>");
        out.print("<a href=" + request.getContextPath() + "/input.html>回到首页</a>");
        out.println(" </BODY>");
        out.println("</HTML>");
        out.flush();
        out.close();
    }
@Override
    protected void doGet(HttpServletRequest request, HttpServletResponse response)
            throws ServletException, IOException {
```

```
            doPost(request, response);
        }
    }
```

(4) 在IE地址栏输入“http://localhost:8080/MySixthWebApp/input.html”，显示如图 6-16 所示的页面。

图 6-16

(5) 输入图书的编号，如输入 ISBN 为“978-7-302”，单击“提交”按钮，显示如图 6-17 所示的页面。

图 6-17

6.6.2　使用 sendRedirect()方法

HttpServletResponse 接口的 sendRedirect(String location)方法可以将响应定向到参数 location 指定的新的 URL。location 可以是一个绝对的 URL，如 response.sendRedirect ("http://java.sun.com")，也可以使用相对的 URL，如果 location 以“/”开头，则容器认为相对于当前 Web 服务器的根，在使用相对的 URL 时，容器将解析为相对于当前的 Web 应用的根。这种重定向的方法，将导致客户端浏览器的请求 URL 跳转，从浏览器的地址栏中可以看到新的 URL 地址，作用类似于上面设置 HTTP 响应头信息的实现。

RequestDispatcher.forward()方法和 HttpServletResponse.sendRedirect()方法的区别是：前者仅是容器中控制权的转向，在客户端浏览器地址栏中不会显示出转向后的地址，例如图 6-17 的地址栏就没有发生变化；后者则是完全的跳转，浏览器将会得到跳转的地址，并重新发送请求链接。这样，从浏览器的地址栏中可以看到跳转后的链接地址。另外，forward()方法可以共享 request 范围内的对象，而 sendRedirect()方法则不行。

现在我们来更改示例 6-19 的代码，取名 GetInfoServlet1，使用 sendRedirect()方法把页面重定向到 DelegateServlet，代码如示例 6-21 所示。

示例 6-21：

```
package control;

import java.io.IOException;
import javax.servlet.ServletException;
import javax.servlet.http.HttpServlet;
import javax.servlet.http.HttpServletRequest;
import javax.servlet.http.HttpServletResponse;

public class GetInfoServlet1 extends HttpServlet {

    public void doGet(HttpServletRequest request,HttpServletResponse response)
        throws ServletException, IOException {
        doPost(request, response);
    }

    public void doPost(HttpServletRequest request,HttpServletResponse response)
        throws ServletException, IOException {
        request.setCharacterEncoding("UTF-8");
        response.setContentType("text/html;charset=UTF-8");
        String bookid = request.getParameter("bookid");
        //把 bookid 存放到 request 对象中
        request.setAttribute("com.control.bookid", bookid);
            //这里没有使用"/"
        response.sendRedirect("delegateServlet?name= bush ");
    }
}
```

同时，我们把示例 6-18 的 input.html 中的 action 指向“getInfoServlet1”，再次运行该页面，输入要查找的图书 ISBN 号，例如“978-7-302”，单击“提交”按钮，将出现如图 6-18 所示的页面。

图 6-18

可以看出，书号是“null”，不是“978-7-302”，原因是，在 DelegateServlet 中，我们并不能获取在 GetInfoServlet1 中放入到 request 对象属性中的 bookid 的值。这就说明了使用 sendRedirect()方法实际上是产生了一个新的请求，所以取不到放在上一个请求中的数据。

总之，不管使用哪种方式来转向，都意味着控制作用，也就是说，Servlet 很多时候是起着控制的作用，所以在 Java Web 的应用中，我们称 Servlet 是控制器(Controller)，由它来实现页面间的转发。在单元九中，我们会详细介绍控制器。

【单元小结】

- 使用 Session 可以实现会话级的数据共享。
- getServletContext()方法用于返回 Servlet 上下文。
- 属于同一个应用程序的 Servlet 使用 RequestDispatcher 对象的 forward()和 include()方法进行通信。
- ServletContext 用于存储不同的 Servlet 的信息，实现 Servlet 间的通信。

【单元自测】

1. 在 Java Web 中，(　　)接口的(　　)方法用于创建会话。

   A. HttpServletRequest、getSession()　　B. HttpServletResponse、newSession()

   C. HtttpSession、newInstance()　　D. HttpSession、getSession()

2. 在 Servlet 中，能实现重定向的方法有(　　)。

   A. 运用 javax.servlet.http.HttpServletRequest 接口的 sendRedirect()方法

   B. 运用 javax.servlet.http.HttpServletResponse 接口的 sendRedirect()方法

   C. 运用 javax.servlet.RequestDispatcher 接口的 forward()方法

   D. 运用 javax.servlet.ResponseDispatcher 接口的 forward()方法

3. 关于 Session 的使用，下列说法正确的是(　　)。

   A. 通过重新运行浏览器程序所打开的不同的用户窗口具有不同的 Session

   B. 通过重新运行浏览器程序所打开的不同的用户窗口具有相同的 Session

   C. Session 可能会超时

   D. Session 永远不可能超时

4. 在 Java Web 的 web.xml 中，有如下代码：<session-config><session-timeout>30</session-timeout></session-config>。上述代码定义了默认的会话超时时长，时长为 30(　　)。

   A. 毫秒　　B. 秒　　C. 分钟　　D. 小时

5. 给定一个 Servlet 的 doGet()方法中的代码片段，如下：request.setAttribute("name","zhang");response.sendRedirect("http://localhost:8080/myServlet");那么在 myServlet 中可以使用(　　)方法把属性 name 的值取出来。

   A. String str=request.getAttribute("name");

   B. String str=(String)request.getAttribute("name");

   C. Object str=request.getAttribute("name");

   D. 无法取出来

【上机实战】

上机目标

- 掌握 ServletContext 对象的使用。
- 在 Web 应用程序中使用会话对象。

上机练习

◆ 第一阶段 ◆

练习 1：使用 ServletContext 记录网站的访客数量

【问题描述】

使用 ServletContext 对象来记录并显示网站的访客数量。

【问题分析】

(1) 要记录站点的访客数量，需要使用服务器端的 ServletContext，因为它的生存周期很长。我们可以把一个计数器存放在 ServletContext 中。

(2) 每次页面被访问，我们让计数器加 1，然后再存回 ServletContext。

【参考步骤】

(1) 创建一个名为 CounterServlet 的 Servlet，完整代码如示例 6-22 所示。

示例 6-22：

```
package control ;
import java.io.IOException;
import java.io.PrintWriter;
import javax.servlet.ServletConfig;
import javax.servlet.ServletContext;
import javax.servlet.ServletException;
import javax.servlet.http.HttpServlet;
import javax.servlet.http.HttpServletRequest;
import javax.servlet.http.HttpServletResponse;
public class CounterServlet extends HttpServlet {
    private ServletContext    ctx    ;
    int counter = 0 ;

    public void doGet(HttpServletRequest request, HttpServletResponse response)
            throws ServletException, IOException {
        doPost(request, response);
    }
```

```
    public void init(ServletConfig conf) throws ServletException {
        super.init(conf);
        ctx =this.getServletContext();
    }

    public void doPost(HttpServletRequest req, HttpServletResponse res)
            throws ServletException, IOException {
        res.setContentType("Text/Html;charset=gb2312");
        PrintWriter out = res.getWriter();
        counter++;
        ctx.setAttribute("counter", new Integer(counter));
        out.println("<HTML>");
        out.println("<HEAD><TITLE>基本 Servlet</TITLE></HEAD>");
        out.println("<H3>欢迎来到我的站点</H3>");
        out.println("<BODY>");
        out.println("您是本网站的第 " + ctx.getAttribute("counter") + " 位访客");
        out.println("</BODY>");
        out.println("</HTML>");
    }
}
```

(2) 发布该Web应用到Tomcat服务器并启动，打开IE，在地址栏中输入“http://localhost:8080/MySixthWebApp/CounterServlet”，显示结果如图6-19所示。当单击“刷新”按钮时，计数器值会增加。如果新打开一个浏览器，可以看到计数器的值不是从1开始，而是延续上次用户访问的值。

图 6-19

◆ 第二阶段 ◆

练习 2：使用 Session 对象模拟网上购物的购物车

【问题分析】

(1) 生成一个List容器对象，把用户选择的书的编号(bookid)放入这个容器对象中，并用HttpSession对象的setAttribute()方法来保存这个List对象。

(2) 在显示购物车的页面，从HttpSession对象中取出存放在HttpSession对象中的List对象，再从List对象中获得用户选择的书的编号(bookid)，然后通过bookid从数据库中取出书的详细信息并显示出来。

【参考步骤】

(1) 建立一个 Web 工程，取名叫作 SessionTraceDemo3。

(2) 创建DBHelper.java和DBCommand.java，以及数据库操作的基础类BaseDAO.java、BookDAO.java和BookDAOImpl.java及业务处理类BookService.java和BookServiceImpl.java。

(3) 创建一个名为 ShowBookServlet 的 Servlet，完整代码如示例 6-23 所示。

示例 6-23：

```
package control;
import java.io.IOException;
import java.io.PrintWriter;
import javax.servlet.ServletException;
import javax.servlet.http.HttpServlet;
import javax.servlet.http.HttpServletRequest;
import javax.servlet.http.HttpServletResponse;
import bean.BookBean;
import java.util.List;
import service.BookService;
import service.BookServiceImpl;
public class ShowBookServlet extends HttpServlet {
    public void doGet(HttpServletRequest request, HttpServletResponse response)
            throws ServletException, IOException {
        doPost(request, response);
    }

    public void doPost(HttpServletRequest request, HttpServletResponse response)
            throws ServletException, IOException {
        request.setCharacterEncoding("UTF-8");
        response.setContentType("text/html;charset=UTF-8");
        //实例化业务类
        BookService service = new BookServiceImpl();
        List<BookBean> list = service.getAllBooks();

        PrintWriter out = response.getWriter();
        out.println("<HTML>");
        out.println("<HEAD><TITLE>取数据库数据</TITLE></HEAD>");
        out.println("<h3>欢迎光临网上书店</h3>");
        out.println("请选择你购买的书，加入到购物车");
         out.println("<BODY><table>");
         out.println("<tr><td>图书编号</td><td>作者</td><td></td>");
        for (int i = 0; i < list.size(); i++) {
            BookBean book = (BookBean) list.get(i);
            out.println("<tr>");
            out.println("<td><a href=showBookDetailServlet?bookid="
                    + book.getBookId()+ ">" + book.getBookId() + "</a></td>");
```

```
                out.println("<td>作者：" + book.getAuthor() + "</td>");
                out.println("<td><a href=addToCartServlet?bookid="
                        + book.getBookId() + ">加入购物车</a></td>");
                out.println("</tr>");
            }
            out.println("</table>");
            out.println("<h4><a href=showShoppingCartServlet>查看购物</a></h4>");
            out.println("</BODY></HTML>");
            out.flush();
            out.close();
        }
    }
```

(4) 创建一个名为 ShowBookDetailServlet 的 Servlet，使用户能够单击图书编号查看到书的详细信息，代码如示例 6-24 所示。

示例 6-24：

```
package control;

import java.io.IOException;
import java.io.PrintWriter;

import javax.servlet.ServletException;
import javax.servlet.http.HttpServlet;
import javax.servlet.http.HttpServletRequest;
import javax.servlet.http.HttpServletResponse;

import service.BookService;
import service.BookServiceImpl;
import bean.BookBean;

public class ShowBookDetailServlet extends HttpServlet {

    public void doGet(HttpServletRequest request,HttpServletResponse response)
            throws ServletException, IOException {
        doPost(request, response);
    }

    public void doPost(HttpServletRequest request,HttpServletResponse response)
            throws ServletException, IOException {
        request.setCharacterEncoding("UTF-8");
        response.setContentType("text/html;charset=UTF-8");
        String bookid=request.getParameter("bookid");
        Integer id=new Integer(bookid);

        //实例业务处理类
        BookService service = new BookServiceImpl();
```

```
            BookBean book=service.getBookById(id);
            PrintWriter out = response.getWriter();
            out.println("<HEAD><TITLE>取数据库数据</TITLE></HEAD>");
            out.println("<h3>图书详细信息如下</h3>");
            out.println("<BODY><table>");
            out.print("<tr><td>图书编号：</td><td>"
                +book.getBookId()+"</td></tr>");
            out.print("<tr><td>ISBN 编号：</td><td>"
                +book.getIsbn()+"</td></tr>");
            out.print("<tr><td>书名：</td><td>"
                +book.getName()+"</td></tr>");
            out.print("<tr><td>作者：</td><td>"
                +book.getAuthor()+"</td></tr>");
            out.print("<tr><td>单价：</td><td>"
                +book.getPrice()+"</td></tr>");
            out.print("<tr><td>库存数量：</td><td>"
                +book.getStock()+"</td></tr>");
            out.print("<tr><td><a href=showBookServlet>回到主页面</a></td><td><a
                href=addToCartServlet?bookid="+book.getBookId()+">添加到购物车</a></td></tr>");
            out.println("</table></BODY>");
            out.println("</HTML>");
            out.flush();
            out.close();
        }
    }
```

(5) 创建一个名为 AddToCartServlet 的 Servlet，用户可以把喜爱的图书加入购物车，代码如示例 6-25 所示。

示例 6-25：

```
package control;
import java.util.List;
import java.io.IOException;
import java.util.ArrayList;
import javax.servlet.RequestDispatcher;
import javax.servlet.ServletException;
import javax.servlet.http.HttpServlet;
import javax.servlet.http.HttpServletRequest;
import javax.servlet.http.HttpServletResponse;
import javax.servlet.http.HttpSession;
public class AddToCartServlet extends HttpServlet {

    public void doGet(HttpServletRequest request,HttpServletResponse response)
            throws ServletException, IOException {
        doPost(request, response);
    }
```

```
    public void doPost(HttpServletRequest request,HttpServletResponse response)
            throws ServletException, IOException {
        request.setCharacterEncoding("UTF-8");
        response.setContentType("text/html;charset=UTF-8");
        String bookid = request.getParameter("bookid");
        HttpSession session = request.getSession();
        List<String> list = null;
        list = (List<String>)session.getAttribute("cart");
        if (list == null) {
            list = new ArrayList<String>();
            session.setAttribute("cart", list);
        }
        list.add(bookid);
        RequestDispatcher rd =request.getRequestDispatcher("showBookServlet");
        rd.forward(request, response);
    }
}
```

(6) 创建一个名为 ShowShoppingCartServlet 的 Servlet，用户可以通过链接查看购物车内的内容，代码如示例 6-26 所示。

示例 6-26：

```
package control;
import java.io.IOException;
import java.io.PrintWriter;
import javax.servlet.RequestDispatcher;
import javax.servlet.ServletException;
import javax.servlet.http.HttpServlet;
import javax.servlet.http.HttpServletRequest;
import javax.servlet.http.HttpServletResponse;
import javax.servlet.http.HttpSession;
import service.BookService;
import service.BookServiceImpl;
import bean.BookBean;
import java.util.List;
public class ShowShoppingCartServlet extends HttpServlet {
    public void doGet(HttpServletRequest request, HttpServletResponse response)
            throws ServletException, IOException {
        doPost(request, response);
    }

    public void doPost(HttpServletRequest request, HttpServletResponse response)
            throws ServletException, IOException {
        request.setCharacterEncoding("UTF-8");
        response.setContentType("text/html;charset=UTF-8");
        HttpSession session = request.getSession();
        List<String> list = (List<String>) session.getAttribute("cart");
```

```
        PrintWriter out = response.getWriter();
        if (list == null) {
            out.print("<h1><font color = red >购物车为空</font></h1>");
            RequestDispatcher rd = request
                    .getRequestDispatcher("showBookServlet");
            rd.include(request, response);
        } else {
            BookService service = new   BookServiceImpl();
            out.println("<HTML>");
            out.println("<HEAD><TITLE>取数据库数据</TITLE></HEAD>");
            out.println(" <h3>欢迎光临网上书店</h3>");
            out.println(" <h3>下面是您已购买的书</h3>");
            out.println("<BODY><table>");
            for (int i = 0; i < list.size(); i++) {
                String index = (String) list.get(i);
                BookBean book = service.getBookById(new Integer(index));
                out.println("<tr>");
                out.println("<td>" + book.getBookId() + "</td>");
                out.println("<td>作者： " + book.getAuthor()+ "</td>");
                out.println("<td>" + book.getPrice()+ "</td>");
                out.println("</tr>");
            }
            out.println("</table>");
            out.println("<h4><a href=showBookServlet>回到主页</a></h4>");
            out.println("</BODY></HTML>");
        }
        out.flush();
        out.close();
    }
}
```

(7) 发布该Web应用到Tomcat服务器并启动，打开IE，在地址栏中输入“http://127.0.0.1:8080/ SessionTraceDemo3/showBookServlet”，显示结果如图 6-20 所示。

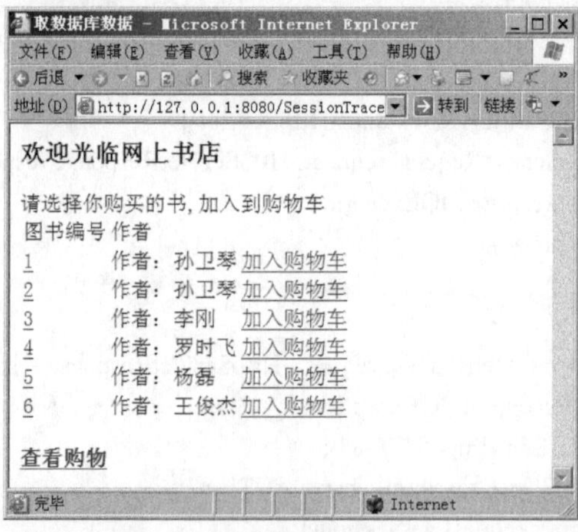

图 6-20

(8) 当用户单击图书编号的超链接时，会显示图书的详细信息，显示结果如图 6-21 所示。

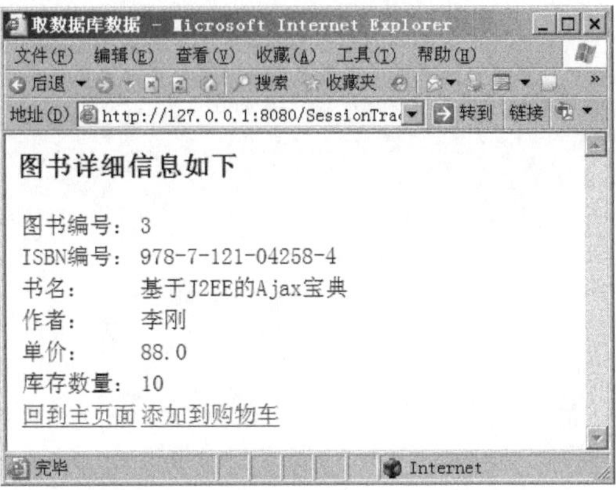

图 6-21

(9) 此时，可以把感兴趣的图书添加到购物车或是直接返回到主页，若选择添加到购物车，此时该书会被放入到购物车，然后页面自动转到主页。可以在主页里继续查看其他书的详细信息或是查看购物车的内容。如果没有在购物车中添加图书，则结果如图 6-22 所示，当添加了若干本书时，结果如图 6-23 所示。

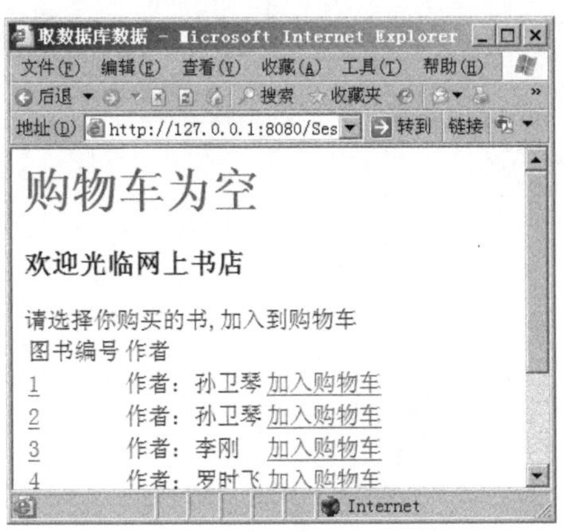

图 6-22

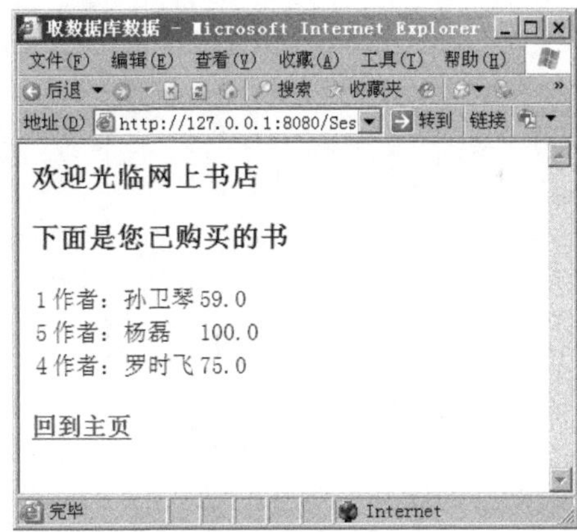

图 6-23

【拓展作业】

1. 第二阶段练习 2 有个小问题，就是当再次选择相同编号的图书时，购物车不能在原来的数量上累加，改进上述问题。

2. 给第二阶段练习 2 的购物车增加删除功能，实现用户可以删除购物车中的书籍。

单元七 表达式语言

课程目标

- 了解什么是表达式语言
- 了解怎么使用表达式语言
- 了解引用隐式对象
- 了解运算符的应用

简 介

EL(Expression Language，表达式语言)是在 JSTL 1.0 规范中引入的，作为 JSTL 的组成部分，当时用来作为 Java 表达式工作，而该表达式必须配合 JSTL 的标签库才能得到需要的结果。在 JSTL 1.1 规范中，JSP 2.0 容器已经能够独立地理解任何 EL 表达式。EL 可以独立出现在 JSP 页面的任何角落。

7.1 应用表达式语言的动力

表达式语言是 JSP 2.0 的新特性，是从 JavaScript 脚本语言中得到启发的一种语言，它借鉴了 JavaScript 多类型转换无关性的特点，在使用 EL 从 scope 中得到参数时可以自动转换类型，因此对于类型的限制更加宽松。Web 服务器对于 request 请求参数通常会以 String 类型来发送，在得到时使用的 Java 语言脚本就应该是 request.getParameter("XXX")，这样的话，对于实际应用还必须进行强制类型转换。而 EL 就将用户从这种类型转换的烦琐工作中脱离出来，允许用户直接使用 EL 表达式取得值，而不用关心它是什么类型。表达式语言可以将表示层简化，它允许我们使用更简短、更易读的项，如下所示：

```
${expression}
```

EL 不仅可以用来取代难以维护的 Java 脚本或笨拙的<jsp:useBean>和<jsp:getProperty>标签，尤其重要的是，表达式语言支持下面的功能。

(1) 精确地访问存储对象：要输出“作用域变量”(使用 setAttribute 方法存储在 PageContext、HttpServletRequest、HttpSession 或 ServletContext 中的对象)，可以直接使用${对象名}。

(2) Bean 属性的简略记法：如果要输出访问作用域变量 company 的 name 属性(即 company.getName()方法的结果)，可以直接使用${company.name}。而要输出访问作用域变量 company 的 stu 属性的 names 属性，则使用${company.stu.names}。

(3) 对集合元素的简单访问：如果要访问数组、List 或 Map 集合中的元素，可使用${对象名[索引]}。如果索引或键所采用的形式可以满足合法 Java 变量名的要求，则 Bean 的点号记法可以和集合的括号记法互换。

(4) 对请求参数、Cookie 和其他请求数据的简单访问：如果要访问标准的请求数据，可以使用几个预定义的隐式对象。

(5) 条件性输出：在进行有选择的输出时，可以不必借助 Java 脚本元素。取而代之，可以使用${test?option1：option2}来完成这一功能。

(6) 自动类型转换：表达式语言移除了大多数类型转换的需求，可以省略很多将字符串解析成数字的代码。

(7) 空值取代错误消息：大多数情况下，没有相应的值或 NullPointerException 异常都会导致空字符串的出现，而非抛出异常。

7.2 EL 语法

EL 表达式语言的语法格式如下。

```
${表达式}
```

所有 EL 都是以${为开始、以}为结尾的。如下面的范例所示，是从 Session 的范围中取得用户的性别。

```
范例：${sessionScope.user.sex}
```

假若依照之前 JSP Scriptlet 的写法，代码如示例 7-1 所示。

示例 7-1：

```
User user =(User)session.getAttribute("user");
String sex =user.getSex( );
```

从中可以发现 EL 的语法比传统的 JSP Scriptlet 更为方便、简洁。下面列举一些合法的 EL 语法格式，如示例 7-2 所示。

示例 7-2：

```
${"hello"}                        //输出字符串常量
${25.5}                           //输出浮点数常量
${25+5}                           //输出算术运算结果
${2>1}                            //输出关系运算结果
${8||6}                           //输出逻辑运算结果
${2>1?2:1}                        //输出条件运算结果
${empty username}                 //输出 empty 运算结果
${username}                       //查找输出变量值
${sessionScope.user.sex}          //输出隐含对象中的属性值
${hopu:fun(arg)}                  //输出自定义函数的返回值
```

7.3 EL 中的常量

EL 表达式中的常量包括布尔常量、整型常量、浮点数常量、字符串常量和 NULL 常量。下面一一讲解。

- 布尔常量，用于区分事物的正反面，即 true 或 false。例如${true}。
- 整型常量，与 Java 中定义的整型常量相同，范围为 Long.MIN_VALUE 到 Long.MAX_VALUE 之间。例如${666}。

- 浮点数常量，与 Java 中定义的浮点数常量相同，范围为 Double.MIN_VALUE 到 Double.MAX_VALUE 之间。例如${666.66}。
- 字符串常量，是用单引号或双引号引起来的一连串字符。例如${"你好！"}。
- NULL 常量，用于表示引用的对象为空，用 null 表示，但是在 EL 表达式中并不会输出 null 而是输出空。例如${null}，页面会什么都不输出。

7.4 EL 中的变量

EL 表达式中的变量不同于 JSP 表达式从当前页面中定义的变量进行查找，而是由 EL 引擎调用 PageContext.findAttribute(String)方法从 JSP 四大作用域范围中查找。例如：${username}，表达式将按照 page、request、session 和 application 范围的顺序依次查找名为 username 的属性；假如中途找到，就直接回传，不再继续找下去；如果全部范围都没有找到，就会返回 null。因此在使用 EL 表达式访问某个变量时，应该指定查找的范围，从而避免在不同作用范围中有同名属性的问题，也提高了查询效率。

EL 中的变量除了要遵循 Java 变量的命名规范外，还需要注意不能使用 EL 中的保留字。EL 中预保留字如下。

and	or	not	empty	div
mod	instanceof	eq	ne	lt
gt	le	ge	true	false
null				

7.5 点号记法与数组记法的等同性

使用表达式语言时，也可以使用数组记法(方括号)来替代点号记法。

```
${mybean.accName}
```

替换成：

```
${mybean ["accName"]}
```

在访问 Bean 的属性时，很少使用第二种形式。

7.6 EL 错误处理机制

作为表现层的 JSP 页面的错误处理，往往会对用户有直观的体验，为此 EL 提供了比较友好的处理方式：不提供警告，只提供默认值和错误，默认值是空字符串，错误是抛出一个异常。EL 对以下几种常见错误的处理方式为：

- 在 EL 中访问一个不存在的变量，则表达式输出空字符串，而不是输出 null；
- 在 EL 中访问一个不存在对象的属性，则表达式输出空字符串，而不会抛出 NullPointerException 异常；
- 在EL中访问一个存在对象的不存在属性，则表达式会抛出PropertyNotFoundException 异常。

7.7 EL 获取数据

EL 表达式主要用于替换 JSP 页面中的脚本表达式，以从各种类型的 Web 域中检索 Java 对象、获取数据。所以使用 EL 获取数据是学习 EL 的一个主要知识点。通过 EL 表达式也可以很轻松地获取 JavaBean 属性或获取数组、Collection、Map 类型集合的数据，下面的小节就从这几个方面进行讲解。

7.7.1 获取 JavaBean 的数据

在使用 EL 获取 JavaBean 的数据时，采用点号记号，EL 会自动去该类中调用对应的 get()方法，如下面示例 7-3 所示。

示例 7-3：

```
<%@ page language="java" contentType="text/html; charset=UTF-8"
    pageEncoding="UTF-8"%>
    <%@ page import="domain.*" %>
<!DOCTYPE html>
<html>
<head>
<meta charset="UTF-8">
<title>获取 javabean 中的属性</title>
</head>
<body>
<h1>获取 javabean 中的属性</h1>
<%
Person p = new Person();
p.setName("习大大");
p.setAge(50);
application.setAttribute("p", p);
%>
${p.name }：${p.age }<!-- el 方式 -->
</body>
</html>
```

7.7.2 获取数组的数据

使用 EL 获取数组的数据也很简单，与 Java 一样，只不过加上了${}而已，如下面示例 7-4 所示。

示例 7-4：

```
<h3>获取数组的数据</h3>
<%
String[]arrs={"math","chinese","english"};
pageContext.setAttribute("arrs", arrs);
%>
${arrs[0]}
${arrs[1]}
${arrs[2]}
```

7.7.3 获取 List 集合的数据

在掌握了获取数组中数据的方法后，学习获取 List 集合的数据也会很容易，如下面示例 7-5 所示。

示例 7-5：

```
<h3>获取 List 集合的数据</h3>
<%
List<String> list=new ArrayList<String>();
list.add("java");
list.add("php");
list.add("web");
pageContext.setAttribute("list", list);
%>
${list[0] }
${list[1] }
${list[2] }
```

7.7.4 获取 Map 集合的数据

使用 EL 获取 Map 集合数据与 list 集合类似，但是如果 Map 的 key 中包含了特殊字符，就不能使用“.”操作而必须使用“[]”操作符。如下面示例 7-6 所示。

示例 7-6：

```
<h3>获取 Map 中的数据</h3>
```

```
<%
Map<String,String> map=new HashMap<String,String>();
map.put("name","tom");
map.put("sex","man");
map.put("address","beijing");
map.put("user.age","18");
pageContext.setAttribute("map", map);
%>
${map.name }
${map.sex }
${map.address }
${map["user.age"] }<!-- 这里不能使用.操作，只能使用[]操作符。 -->
```

7.8　EL 内置对象

JSP 开发人员可以在 EL 表达式中使用 EL 内置对象，而无须任何显式地声明编码。内置对象就是设计用来为 JSP 编程提供方便的，通过内置对象，用 EL 编写的代码就可以直接使用 JSP 页面中的一些最常用的地方。总共有 11 个内置对象，这 11 个内置对象可以分为如下 5 大类：

- JSP 内置对象；
- 作用域访问内置对象；
- 参数访问内置对象；
- HTTP 头访问内置对象；
- 初始化参数访问内置对象。

各类别的对象和作用描述如表 7-1 所示。

表 7-1

类　别	对象标识符	作　用
JSP	pageContext	与页面相对应的页面上下文对象
作用域	pageScope	页面作用域中属性/值的集合(Map)
	requestScope	请求作用域中属性/值的集合(Map)
	sessionScope	会话作用域中属性/值的集合(Map)
	applicationScope	应用程序作用域中属性/值的集合(Map)
请求参数	param	请求参数/单值字符串的集合(Map)
	paramValues	请求参数的所有值作为 String 数组存储的集合(Map)
请求头	header	请求头名称/单值字符串的集合(Map)
	headerValues	请求头名称/多值作为 String 数组存储的集合(Map)
Cookie	cookie	所有 cookie 组成的集合(Map)
初始化参数	initParam	Web 应用程序上下文初始化参数的集合(Map)

EL pageContext 内置对象是 javax.servlet.jsp.PageContext 类的实例，容器自动为每个 JSP 页面创建与之对应的 pageContext 对象，并把与 JSP 对应的 Servlet 对象的相关对象都自动地加入到 pageContext 对象中，它与同名的 JSP 内置对象实际上就是同一个对象。余下的 EL 内置对象都是 java.util.Map 类，它们只是提供了更容易的途径来访问 pageContext 内置对象的某些性质。

4 个作用域内置对象，使用它们可以很容易地访问作用域属性。例如：存储到 request 作用域的 username 属性可以通过 EL 表达式${ requestScope.username}来直接访问。

两个参数访问内置对象，可以用来访问 HTTP 请求参数(表单提交参数)，即 param 和 paramValues。param 是一个用于访问单值参数的映射，paramValues 则可用于访问可能包含多个值的参数。如果当前请求中不存在这个参数，则返回空字符串，而非 null。

3 个 HTTP 请求头访问内置对象，可以用于访问 HTTP 请求头，分别是 header、headerValues 和 cookie。

初始化参数访问内置对象 initParam，这个映射可以用于访问初始化参数的值，初始化参数的值一般都在 web.xml 中设置。

下面看一个使用 EL 内置对象的综合例子，示例 7-7 是一个表单，在单元五示例 5-1 中使用过，该页面提交到 showInfo.jsp 页面，和单元五示例 5-2 不同的是，在示例 7-8 的页面中使用 EL 内置对象。

示例 7-7：

```
<%@ page language="java" contentType="text/html; charset=UTF-8"
    pageEncoding="UTF-8"%>
<!DOCTYPE html>
<html>
<head>
<meta charset="UTF-8">
<title>银行账户注册页面</title>
</head>
<body>
 <form action="showInfo.jsp" method="post"> 
    请输入你的注册信息：  <br>
    账号：<input type="text" name="accId"><br>
    名字：<input type="text" name="accName"><br>
    密码：<input type="password" name="accPwd"><br>
    开户金额:<input type="text" name="balance"><br>
    <input type="submit" value="注册">
 </form>
</body>
</html>
```

示例 7-8：

```
<%@ page language="java" contentType="text/html; charset=UTF-8"
```

```
    pageEncoding="UTF-8"%>
<!DOCTYPE html>
<html>
<head>
<meta charset="UTF-8">
<title>显示注册信息</title>
</head>
<body>
 <%request.setCharacterEncoding("UTF-8"); %>
    请确认以下的注册信息:<br>
    账号：${param.accId}<br>
    名字：${param.accName}<br>
    密码：${param.accPwd}<br>
    开户金额：${param.balance}<br>
</body>
</html>
```

7.9　EL 中的运算符

1. 算术运算符

算术运算符包括以下 3 项。

- +和-：它们是常规的加法和减法运算符。
- *和/：它们是常规的乘法和除法运算符。
- %和 mod：%(或与之等同的 mod)运算符用来计算模数(余数)，与 Java 编程语言中的%相同。

2. 关系运算符

关系运算符包括以下 3 项。

- ==和 eq：这两个相等性运算符检查参数是否相等。
- !=和 ne：这两个相等性运算符检查参数是否不同。
- <和 lt，>和 gt，<=和 le，>=和 ge。

3. 逻辑运算符

逻辑运算符用来组合关系运算符得出结果，有&&、and、||、or、!和 not。

4. 空运算符

空运算符为 empty，如果这个运算符的参数为 null、空字符串、空数组、空 Map 或空集合，则返回 true，否则返回 false。

表 7-2 列出了表达式使用 EL 运算符的计算结果。

表 7-2

表达式	计算结果
${3+2}	5
${3-2}	1
${3*2}	6
${3/2}	1.5
${3 mod 2}	1
${3<2} or ${3 lt 2}	false
${3>2} or ${3 gt 2}	true
${3.2>=2.2} or ${3.2 ge 2.2}	true
${3 >2 ? 3: 2 }	3
${50.0 == 50 }	true
${10 * 10 ne 100 }	false
${' c ' < ' d ' }	false
${'tim ' gt ' tin ' }	false
${3.2 E 4 +5.4}	32005.4
${1 div 4}	0.75

【单元小结】

- 停用个别页面中的表达式语言。
- 访问作用域变量。
- 访问 bean 的属性。
- 点号记法与数组记法的等同性。
- 引用隐式对象。
- 运算符的应用。

【单元自测】

1. JSP 表达式语言用于访问(　　)中存储的数据。

A. JavaBean　　B. Applet　　C. Servlet　　D. Java 程序

2. pageContext 对象提供对(　　)和 request 对象的访问权限。

A. session　　B. param　　C. header　　D. cookie

3. JSP EL 表达式的语法为(　　)。

A. !JSP expression　　B. ${JSP expression}

C. #{ JSP expression }　　D. @{ JSP expression }

4. (　　)运算符用于检查空值。

A. +　　B. %　　C. eq　　D. empty

5. empty 运算符用于(　　)。

A. 检查变量的值　　B. 清除变量的值　　C. A 和 B　　D. 以上都不是

【上机实战】

上机目标

掌握如何在 JSP 页面中使用表达式语言。

上机练习

◆ 第一阶段 ◆

练习 1：使用 EL 表达式改变页面背景和表格的属性

【问题描述】

编写一个 JSP 程序，根据用户输入的属性值来改变当前页面的背景色、页面字体及表格的宽度和边框。

【问题分析】

创建一个表单，用于输入当前页面的背景色、页面字号的值、表格行的背景和表格边框的值。将表单的 action 属性指定为当前页面，并使用 EL 来获取这些属性的值，来改变页面的属性。

【参考步骤】

(1) 创建 JSP 页面 change.jsp，如示例 7-9 所示。

示例 7-9：

```
<%@ page language="java" contentType="text/html; charset=UTF-8"
    pageEncoding="UTF-8"%>
<!DOCTYPE html>
<html>
<head>
<meta charset="UTF-8">
<title>改变页面的属性</title>
</head>
<body bgcolor="${param.bgcolor}">
使用 EL 改变页面属性:
```

```
        <form action="change.jsp" method="post">
        <table border="${param.bsize}">
        <tr>
        <td><font size ="${param.fsize}">页面的背景色：</font></td>
        <td><input type="text" name="bgcolor" value="${param.bgcolor}"/></td>
        </tr>
        <tr>
        <td><font size ="${param.fsize}">页面的字体大小：</font></td>
        <td><input type="text" name="fsize" value="${param.fsize}"></td>
        </tr>
        <tr>
        <td><font size ="${param.fsize}">表格的宽度：</font></td>
        <td><input type="text" name="bwidth" value="${param.bwidth }"></td>
        </tr>
        <tr>
        <td><font size ="${param.fsize}">表格的表框：</font></td>
        <td><input type="text" name="bsize" value="${param.bsize }"></td>
        </tr>
        <tr>
        <td colspan="2" align="center"><input type="submit" value="改变属性"></td>
        </tr>
        </table>
        </form>

    </body>
    </html>
```

(2) 在文本框中输入对应的数值，然后单击“改变属性”按钮，页面元素的属性会随着用户输入的不同而不同。

练习 2：使用表达式语言执行数学运算

【问题描述】

编写一个 JSP 程序，用来模拟 6 位评委分别给歌手打分，分数在 100 分内，并在页面上显示总分和平均分。

【问题分析】

创建一个表单，用以分别输入 6 位评委给同一名歌手的分数。将表单的 action 属性指定为当前页面，从而可将运算的结果输出到同一个 JSP 页面。使用 JSP 表达式语言计算所输入分数的总分和平均分。

【参考步骤】

(1) 创建 JSP 页面，写入如示例 7-10 所示的代码。

示例 7-10：

```
<%@page contentType="text/html;charset=gbk"%>
<html>
    <head><title>计算总分和平均分</title></head>
    <body>
    <form action="score.jsp" method="post">
    <table>
        <tr>
        <td>1 号评委分数：</td>
        <td><input type="text" value="${param.no1}" name="no1" size="5" /></td>
        </tr>
        <tr>
        <td>2 号评委分数：</td>
        <td><input type="text" value="${param.no2}" name="no2" size="5" /></td>
        </tr>
        <tr>
        <td>3 号评委分数：</td>
        <td><input type="text" value="${param.no3}" name="no3" size="5" /></td>
        </tr>
        <tr>
        <td>4 号评委分数：</td>
        <td><input type="text" value="${param.no4}" name="no4" size="5" /></td>
        </tr>
        <tr>
        <td>5 号评委分数：</td>
        <td><input type="text" value="${param.no5}" name="no5" size="5" /></td>
        </tr>
        <tr>
        <td>6 号评委分数：</td>
        <td><input type="text" value="${param.no6}" name="no6" size="5" /></td>
        </tr>
        <tr><td>总分：</td>
        <td>${param.no1 + param.no2 + param.no3 + param.no4+param.no5+param.no6}</td>
        </tr>
        <tr><td>平均分：</td>
        <td>${(param.no1 + param.no2 + param.no3 + param.no4+param.no5+param.no6)/6}</td>
        </tr>
        <tr>
        <td></td>
        <td><input type="submit" value="提交" /></td>
        </tr>
            </table>
        </form>
    </body>
</html>
```

(2) 分别输入 6 位评委的分数，然后单击“提交”按钮。在同一个 JSP 页面上显示总分和平均分，结果如图 7-1 所示。

图 7-1

◆ 第二阶段 ◆

练习 3：改进练习 2，使评分系统具有去掉一个最高分和一个最低分的功能。

【问题分析】

(1) 练习 2 中对 6 个分数计算总分和平均分，本例中要求去掉 6 个分数中的一个最高分和一个最低分。可以声明一个方法，入参是 6 个 String 类型的分数，返回值是一个数组，数组的第一元素是 4 个有效分的总分，第二个元素是 4 个有效分的平均分。

(2) 在页面上输出数组的第一个元素的值(总分)和第二个元素的值(平均分)。

【拓展作业】

1. 编写一个使用 EL 表达式语言的 JSP 程序，输入一个数字，提交后输出该数字的阶乘。

2. 编写一个 JSP 程序，使用 EL 表达式语言显示一个页面被单击的总次数。

单元八

JSP 标准标签库

课程目标

- 了解什么是 JSTL
- 核心标签库的使用
- 国际化标签的使用
- 格式化标签的使用

简 介

JSTL(Java Server pages Standard Tag Library，JSP 标准标签库)是由 SUN 公司推出的，由 Apache Jakarta 组织负责维护的用于编写和开发 JSP 页面的一组标准标签。从 JSP 1.1 规范开始，JSP 就支持使用自定义标签了。由于自定义标签的广泛使用造成了程序员的重复定义，这样就促成了 JSTL 的诞生。作为开源的标准技术，它一直在不断地完善。JSTL 的发布包有两个版本: Standard-1.0 Taglib、Standard-1.1 Taglib。其中，JSTL 1.0 支持 Servlet 2.3 和 JSP 1.2 规范，Tomcat 4.x 支持这些规范。JSTL 1.1 支持 Servlet 2.4 和 JSP 2.0 规范，Tomcat 5.x 或以上版本支持这些规范。

JSTL 作为一个标准的已制定好的标签库，可以应用于各种领域，如基本输入输出、流程控制、循环、XML 文件解析、数据库查询及国际化和格式化的应用等。JSTL 所提供的标签函数库主要分为五大类：核心标签库(Core tag library)、I18N 格式标签库(I18N-capable formatting tag library)、SQL 标签库(SQL tag library)、XML 标签库(XML tag library)和函数标签库(Functions tag library)。

8.1 基本概念

为了更好地理解 JSTL 和自定义标签，首先要明确本单元会使用到的几个概念。

1. 标签

标签(Tag)是一种 XML 元素，通过标签可以使 JSP 网页变得简洁并且易于维护，还可以方便地实现同一个 JSP 文件支持多种语言版本。由于标签是 XML 元素，所以它的名称和属性都是对大小写敏感的。

2. 标签库

由一系列功能相似、逻辑上互相联系的标签构成的集合称为标签库(Tag library)。

3. 标签库描述文件

标签库描述文件(Tag Library Descriptor)是一个 XML 文件，这个文件提供了标签库中类和 JSP 中对标签引用的映射关系。它是一个配置文件，与 web.xml 类似，一般以.tld 作为文件的扩展名。

4. 标签处理类

标签处理类(Tag Handle Class)是一个 Java 类，这个类继承了 TagSupport 或者扩展了 SimpleTag 接口，通过这个类可以实现自定义 JSP 标签的具体功能。

8.2 JSTL 函数库分类

JSTL 由 5 个不同功能的标签库组成，在 JSTL 规范中为这 5 个标签库分别指定了不同的 URI，并对标签库的前缀做出了约定，如表 8-1 所示。

表 8-1

标签库	前置名称	URI	示例
核心标签库	c	http://java.sun.com/jsp/jstl/core	<c:out>
I18N 标签库	fmt	http://java.sun.com/jsp/jstl/fmt	<fmt:formatDate>
SQL 标签库	sql	http://java.sun.com/jsp/jstl/sql	<sql:query>
XML 标签库	x	http://java.sun.com/jsp/jstl/xml	<x:forBach>
函数标签库	fn	http://java.sun.com/jsp/jstl/functions	<fn:split>

核心标签库中包含实现 Web 应用的通用操作的标签。例如，输出变量内容的<c:out>标签、用于条件判断的<c:if>标签和用于循环遍历的<c:forEach>标签等。

I18N 标签库中包含实现 Web 应用程序的国际化标签。例如，设置 JSP 页面的本地信息、设置 JSP 页面的时区和本地敏感的数据(如数值、日期)按照 JSP 页面中设置的本地格式进行显示等。

SQL 标签库中包含用于访问数据库和对数据库中的数据进行操作的标签。例如，从数据源中获得数据库连接、从数据库表中检索数据等。由于在实际开发中，多数应用采用分层开发模式，JSP 页面通常仅用作表现层，并不会在 JSP 页面中直接操作数据库，所以此标签库在分层的较大项目中较少使用，在小型不分层的项目中可以通过 SQL 标签库实现快速开发。

XML 标签库中包含对 XML 文档中的数据进行操作的标签。例如，解析 XML 文档、输出 XML 文档中的内容以及迭代处理 XML 文档中的元素等。

函数标签库由 JSTL 提供一套 EL 自定义函数，包含了 JSP 页面制作者经常要用到的字符串操作，例如，提取字符串中的字符串、获取字符串的长度和处理字符串中的空格等。

由于 SQL 标签库、XML 标签库和函数标签库在实际运用中并不广泛，因此后面小节重点讲解核心标签库和 I18N 标签库。

8.3 JSTL 的安装使用

目前 JSTL 最新版本为 1.2.5，需要在 Servlet 2.5、JSP 2.1 以上环境中运行。JSTL 与所需环境的版本对应关系如图 8-1 所示。

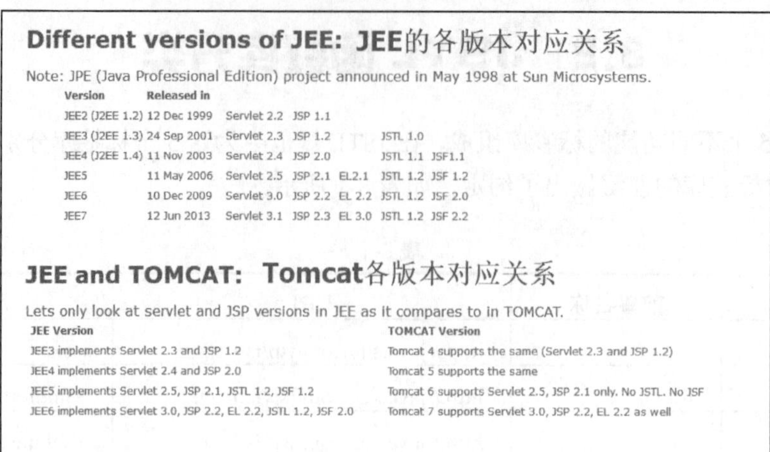

Different versions of JEE: JEE的各版本对应关系

Note: JPE (Java Professional Edition) project announced in May 1998 at Sun Microsystems.

Version	Released in					
JEE2 (J2EE 1.2)	12 Dec 1999	Servlet 2.2	JSP 1.1			
JEE3 (J2EE 1.3)	24 Sep 2001	Servlet 2.3	JSP 1.2		JSTL 1.0	
JEE4 (J2EE 1.4)	11 Nov 2003	Servlet 2.4	JSP 2.0		JSTL 1.1	JSF1.1
JEE5	11 May 2006	Servlet 2.5	JSP 2.1	EL2.1	JSTL 1.2	JSF 1.2
JEE6	10 Dec 2009	Servlet 3.0	JSP 2.2	EL 2.2	JSTL 1.2	JSF 2.0
JEE7	12 Jun 2013	Servlet 3.1	JSP 2.3	EL 3.0	JSTL 1.2	JSF 2.2

JEE and TOMCAT: Tomcat各版本对应关系

Lets only look at servlet and JSP versions in JEE as it compares to in TOMCAT.

JEE Version	TOMCAT Version
JEE3 implements Servlet 2.3 and JSP 1.2	Tomcat 4 supports the same (Servlet 2.3 and JSP 1.2)
JEE4 implements Servlet 2.4 and JSP 2.0	Tomcat 5 supports the same
JEE5 implements Servlet 2.5, JSP 2.1, JSTL 1.2, JSF 1.2	Tomcat 6 supports Servlet 2.5, JSP 2.1 only. No JSTL. No JSF
JEE6 implements Servlet 3.0, JSP 2.2, EL 2.2, JSTL 1.2, JSF 2.0	Tomcat 7 supports Servlet 3.0, JSP 2.2, EL 2.2 as well

图 8-1

在安装使用 JSTL 前，先下载 JSTL 标签库的 jar 包，官方下载地址为：http://tomcat.apache.org/download-taglibs.cgi，下载页面如图 8-2 所示。

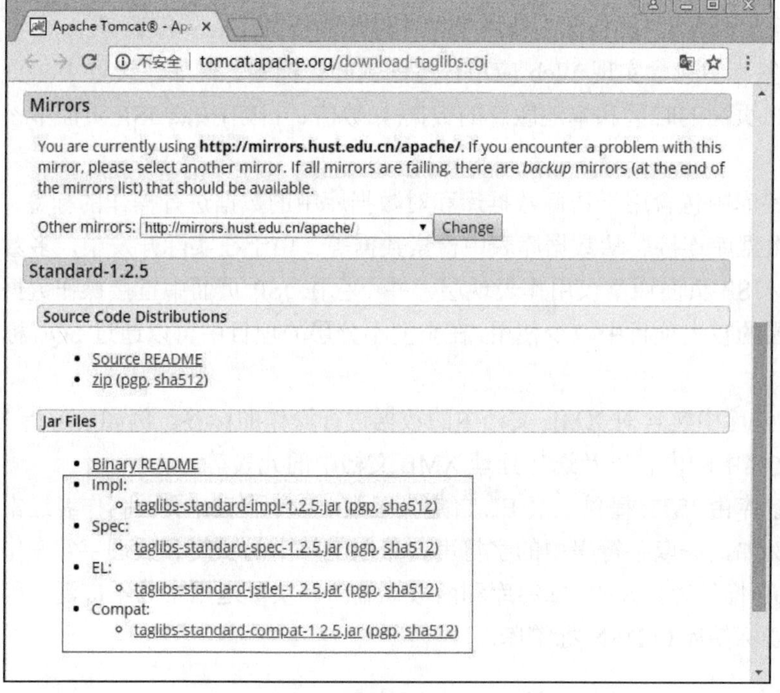

图 8-2

将图 8-2 中下载的 4 个 jar 包复制到 WebContent\WEB-INF\lib 目录下即可使用，效果如图 8-3 所示。

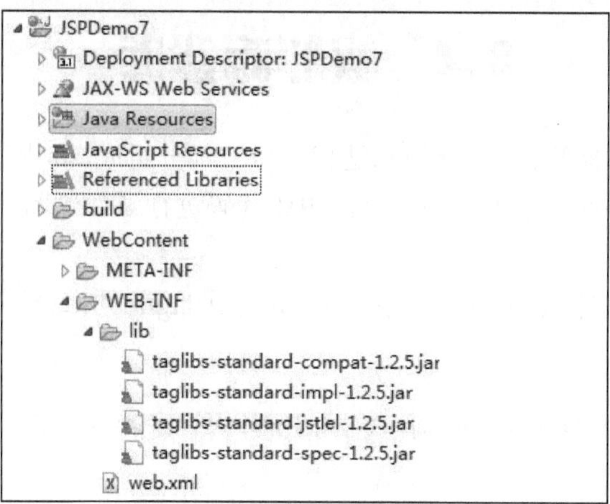

图 8-3

下面用代码演示 JSTL 的使用，如示例 8-1 所示。

示例 8-1：

```
<%@ page language="java" contentType="text/html; charset=UTF-8"
    pageEncoding="UTF-8"%>
    <%@taglib prefix="c" uri="http://java.sun.com/jsp/jstl/core" %>
<!DOCTYPE html>
<html>
<head>
<meta charset="UTF-8">
<title>JSTL 使用示例</title>
</head>
<body>
<c:set var="str" value="Hello JSTL!"/>
<c:out value="${str }"/>
</body>
</html>
```

启动服务，在浏览器输入框内输入 http://localhost:8080/JSPDemo7/hello.jsp，运行结果如图 8-4 所示。

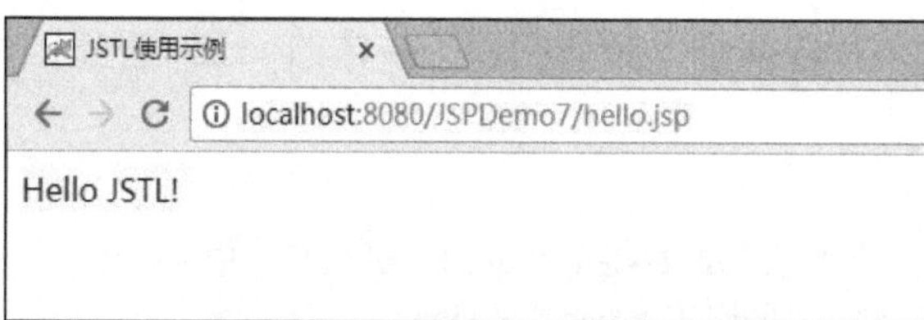

图 8-4

8.4 核心标签库

核心标签库由通用标签、条件标签和迭代标签三部分组成，其中通用标签用于操作 JSP 页面范围的变量，条件标签用于对 JSP 页面中的代码进行条件判断和处理，迭代标签用于循环遍历对象集合。

在 JSP 页面中使用核心标签库，首先需要使用 taglib 指令导入核心标签库的 uri，如下：

```
<%@ taglib uri= " http://java.sun.com/jsp/jstl/core " prefix= "c" %>
```

8.4.1 通用标签

通用标签用于在 JSP 页面中创建、删除变量和显示变量的值，常用的有 3 个，分别是<c:set>、<c:out>和<c:remove>标签。

<c:set>标签用于在某个作用域(page、request、session 和 application)中设置某个变量。如果在这个指定的作用域中，变量不存在，则创建它，并给这个变量赋初值，如果变量已经存在，则把新值赋给这个变量。其语法如下，“[ ]”中的参数表示可选择。

```
<c:set var= " varName " value= " value " [scope= "page|request|session|application"]
 [target= " target "] [property= " propertyName "]/>
```

其中参数的解释如表 8-2 所示。

表 8-2

参　数	作　用
var	指定变量的名称
value	变量的值或表达式的值
scope	指定变量的存储作用域，默认范围为 page
target	指定将要设置属性的实例对象名，它必须是 JavaBean 或者 java.util.Map 对象
property	target 实例对象的属性名称

示例 8-2 所示的代码片段演示了带体的和不带体的<c:set>标签的用法。

示例 8-2：

```
<!--将 level 4 这个值存储到范围为 session 的 userLevel 变量之中-->
<c:set var="userLevel" value="level 4" scope="session"/>

<!--将 JavaBean 的属性值存储至范围为 pageScope 的 cid 变量之中-->
<c:set var="cid" value="${acc.accId}"/>
<!--将标签体里面的值存储至范围为 sessionScope 的 users 变量之中-->
```

```
<c:set var="users" scope="session">
  bush, tom, jack
</c:set>
```

注意，如果 value 的值为 null，var 指定的变量将被删除。如果 var 指定的变量不存在，则会创建一个变量，但仅当 value 的值不为 null 时才会创建新变量。

示例 8-3 的代码片段演示了<c:set>标签的 target、property 用法，其中 acc 和 acc1 是两个 Account 类的 JavaBean 的实例对象。

示例 8-3：

```
<!--将 value 的值存储至范围为 sessionScope 的名为 acc 的 JavaBean 实例的 accName 属性之中-->
<c:set target="${acc}" property="accName" value="猪八戒" scope="session"/>

<!--将标签体里面的值存储至范围为 pageScope 的名为 acc1 的 JavaBean 实例的 accName 属性之中-->
<c:set target="${acc1}" property="accName">
  ${acc.accName}
</c:set>
```

这里需要注意的是，如果 target 是一个 Map 集合，property 指定的是该 Map 的一个键；如果 target 是一个 Bean 实例对象，property 指定的是该 Bean 的一个属性字段；如果 target 表达式值为 null，容器会抛出一个异常；如果 target 表达式不是一个 Map 集合或 Bean 实例对象，容器会抛出一个异常。如果 target 表达式是一个 Bean 实例对象，但是这个 Bean 没有与 property 匹配的属性字段，容器会抛出一个异常。

另外还要注意，不能同时使用 var 和 target 属性。scope 是可选的，如果没有使用这个属性，则默认为页面作用域。当没有使用这个属性时，容器会依次在 pageScope、requestScope、sessionScope、applicationScope 中查找，如果找不到 var 指定名字的属性，容器就会在 pageScope 中新建一个属性；如果找不到 target 指定的对象，容器会抛出一个异常。

<c:remove>用于删除某个变量或属性，其语法为：

```
<c:remove var=" varName " [scope="page|request|session|application"]/>
```

其中参数 var 是待删除的变量的名字，scope 是这个变量的范围。

<c:out>把计算的结果输出到 JspWriter 对象。

```
<c:out value=" value " [escapeXml="true|false"] [default="defaultValue"]/>
```

其中参数的解释如表 8-3 所示。

表 8-3

参　数	作　用
value	将要计算的表达式或变量的值
escapeXml	确定字符<、>、&、'和"在结果字符串中是否被转换成字符实体代码，默认为 true
default	如果 value 是 null，那么输出指定的默认值

escapeXml 字符实体代码表的对应关系如表 8-4 所示。

表 8-4

字　符	字符实体代码
<	<
>	>
&	&
'	'
"	"

下面给出一个综合使用 3 个标签的例子，代码如示例 8-4 所示。

示例 8-4：

```
<%@ page contentType="text/html; charset=gbk" import="bean.Account"%>
<%@ taglib prefix="c" uri="http://java.sun.com/jsp/jstl/core"%>
<html>
    <head><title>核心标签库</title></head>
    <body>
      <jsp:useBean id="acc" class="bean.Account" scope="session"/>
      <jsp:setProperty name="acc" property="accId" value="95599"/>
      <c:set var="userLevel" value="level 4"/>
      <c:set var="cid" value="${acc.accId}"/>
      <c:set var="users">
        bush, tom, jack
        </c:set>
        输出 1:<br>
        userLevel= <c:out value="${pageScope.userLevel}"/><br>
        cid= <c:out value="${cid}"/><br>
        users= <c:out value="${users}"/><br>
        <c:remove var="userLevel"/>
        <c:remove var="cid"/>
        <c:remove var="users"/>
        删除后，输出 1:<br>
        userLevel= <c:out value="${pageScope.userLevel}"/><br>
        cid= <c:out value="${cid}"/><br>
        users= <c:out value="${users}"/><br>

        <jsp:useBean id="acc1" class="bean.Account" scope="session"/>
      <c:set target="${acc}" property="accName" value="猪八戒"/>
      <c:set target="${acc1}" property="accName">
            ${acc.accName}和猪小戒
        </c:set>
            输出 2:<br>
        acc.accName=<c:out value="${acc.accName}"/> <br>
        acc1.accName=<c:out value="${acc1.accName}"/> <br>
```

```
    </body>
</html>
```

结果如图 8-5 所示。

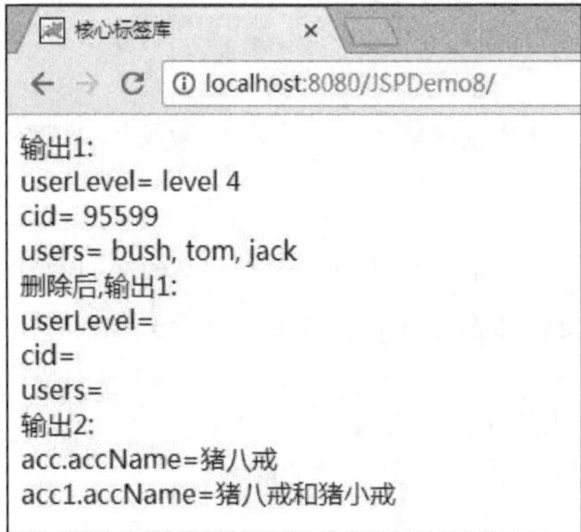

图 8-5

8.4.2　条件标签

条件标签用于支持 JSP 页面中的各种条件。常用的标签有 2 个，分别是<c:if>和<c:choose>。

<c:if>用于进行条件判断，如果它的 test 属性为 true，那么就计算标签体的内容。如果为 false，就忽略标签体的内容。标签体内容除了能放纯文字外，还可以放任何 JSP 程序代码(Scriptlet)、JSP 标签或者 HTML 代码。语法如下：

```
<c:if test=" testCondition " var="varName" [scope=" page|request|session|application "]>
标签体内容
</c:if>
```

属性说明：test 是条件表达式。var 用来指定存放本次条件表达式计算后的值变量，它的类型为 Boolean。scope 是 var 指定变量的作用域。

<c:choose>标签的作用类似于 Java 语言的 switch 语句，它用于执行条件语句块。<c:choose>标签处理<c:when>标签的标签体，可以将多个<c:when>标签嵌入一个<c:choose>标签中。如果条件的值都为 false，则将处理<c:otherwise>的标签体。代码如示例 8-5 所示。

示例 8-5：

```
<%@ page contentType="text/html; charset=gbk" %>
<%@ taglib prefix="c" uri="http://java.sun.com/jstl/core"%>
<jsp:useBean id="acc" class="bean.Account" scope="session"/>
<c:set value="${param.accId}" target="${acc}" property="accId" />
<c:set value="${param.accPwd}" target="${acc}" property="accPwd" />
<c:set value="${param.balance}" target="${acc}" property="balance" />
```

```
<html>
 <body>
 <c:if test="${acc.accId =='95599'   && acc.accPwd==' 888'}" var = "result" scope= "page"/>
   <c:if value="${result}">
   欢迎你！ <c:out value ="acc.accName"/>
  </c:if >
<c:if value="${!result}">
   你的卡号或密码不正确，请重新输入！
  </c:if >
<c:choose>
  <c:when test="${acc.balance < 0}">
    <font color="red">开户金额不能是负数!</font>
  </c:when>
  <c:when test="${ user.age>=0&&acc.balance <10 }">
    <font color="red">开户金额至少是 10 元!</font>
  </c:when>
  <c:otherwise>
    <font color="green">开户金额 OK!</font>
  </c:otherwise>
</c:choose>
</body>
</html>
```

8.4.3 迭代标签

迭代标签用于重复计算标签的标签体。迭代标签有两种，分别是<c:forEach>和<c:forTokens>。

<c:forEach>标签的作用是遍历集合中的元素或者是有条件地重复计算标签体的内容，语法如下。

```
<c:forEach [var=" varName "] items=" collection " [varStatus=" varStatusName "]
  [begin=" begin "] [end=" end "] [step=" step "]>
 标签体内容
</c:forEach>
```

其中参数的解释如表 8-5 所示。

表 8-5

参　数	作　用
var	指定迭代的参数的名字
items	将要迭代的 items 的集合
varStatus	表示迭代的状态，可以访问迭代自身的信息

(续表)

参　数	作　用
begin	如果指定 items 属性，就从 items [begin]开始迭代；如果没有指定 items 属性，就从指定的 begin 值开始循环，相当于 for(int i=begin;;)语句
end	如果指定 items，迭代到 items [end]处结束；如果没有指定 items，那么循环到 end 值处结束，相当于 for(;i<end;)语句
step	迭代的步长

在本单元中，学习了 EL 的隐式对象 header 和 cookie，这两个对象都属于 java.util.Map 集合。示例 8-6 的代码演示了如何使用迭代标签来遍历集合中的元素。

示例 8-6：

```
<%@ page language="java" contentType="text/html; charset=UTF-8"
      pageEncoding="UTF-8"%>
      <%@taglib prefix="c" uri="http://java.sun.com/jsp/jstl/core" %>
<!DOCTYPE html>
<html>
<head>
<meta charset="UTF-8">
<title>显示请求头的内容和 Cookie 内容</title>
</head>
<body>
   读取请求头的信息<br>
      <table border="1">
      <tr bgcolor="green"><td>请求头</td><td>请求名</td></tr>
        <c:forEach items="${header}" var="head" varStatus="status">
        <c:if test="${status.count % 2 !=0}">
          <tr bgcolor="pink"> <td>${head.key}</td><td>${head.value}</td></tr>
        </c:if>
        <c:if test="${status.count % 2 ==0}">
          <tr bgcolor="white"> <td>${head.key}</td><td>${head.value}</td></tr>
        </c:if>
        </c:forEach>
      </table><br>
      读取 cookie 里的信息 <br>
      <table border="1" width="300">
      <tr bgcolor="green"><td>cookie 名</td><td>cookie 值</td></tr>
        <c:forEach items="${cookie}" var="cook" varStatus="st">
        <c:choose>
        <c:when test="${st.count % 2 != 0}">
          <tr bgcolor="pink"> <td>${cook.key}</td><td>${cook.value}</td></tr>
        </c:when>
        <c:otherwise>
        <tr bgcolor="white"> <td>${cook.key}</td><td>${cook.value}</td></tr>
        </c:otherwise>
        </c:choose>
```

```
        </c:forEach>
    </table>
</body>
</html>
```

运行结果如图 8-6 所示。

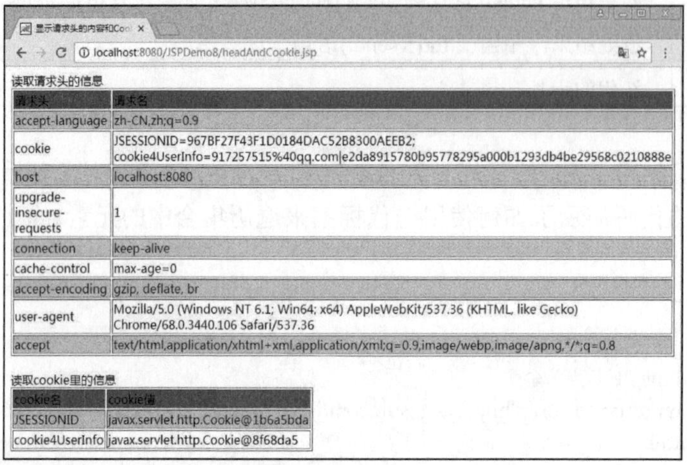

显示请求头的内容和Coo

localhost:8080/JSPDemo8/headAndCookie.jsp

读取请求头的信息

请求头	请求名
accept-language	zh-CN,zh;q=0.9
cookie	JSESSIONID=967BF27F43F1D0184DAC52B8300AEEB2; cookie4UserInfo=917257515%40qq.com|e2da8915780b95778295a000b1293db4be29568c0210888e
host	localhost:8080
upgrade-insecure-requests	1
connection	keep-alive
cache-control	max-age=0
accept-encoding	gzip, deflate, br
user-agent	Mozilla/5.0 (Windows NT 6.1; Win64; x64) AppleWebKit/537.36 (KHTML, like Gecko) Chrome/68.0.3440.106 Safari/537.36
accept	text/html,application/xhtml+xml,application/xml;q=0.9,image/webp,image/apng,*/*;q=0.8

读取cookie里的信息

cookie名	cookie值
JSESSIONID	javax.servlet.http.Cookie@1b6a5bda
cookie4UserInfo	javax.servlet.http.Cookie@8f68da5

图 8-6

<c:forEach>标签还有另外的用法就是重复做标签体内指定的事情，相当于 for 或 while 循环。此时，不需要指定属性 items。示例 8-7 的代码演示了根据步长循环输出数。如果不指定迭代的步长，那么默认 step 的值为 1。

示例 8-7：

```
<%@ page language="java" contentType="text/html; charset=UTF-8"
    pageEncoding="UTF-8"%>
    <%@taglib prefix="c" uri="http://java.sun.com/jsp/jstl/core" %>
<!DOCTYPE html>
<html>
<head>
<meta charset="UTF-8">
<title>循环输出数</title>
</head>
<body>
<h4>循环输出数，步长为 1</h4>
<c:forEach var="i" begin="1" end="10">
  <c:out value="${i}"/>  
</c:forEach>
<h4>循环输出数，步长为 2</h4>
<c:forEach var="i" begin="1" end="10" step="2">
  <c:out value="${i}"/>  
</c:forEach>
</body>
</html>
```

运行结果如图 8-7 所示。

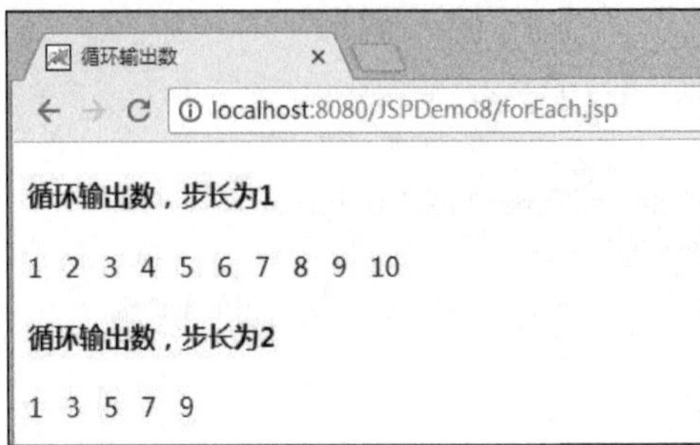

图 8-7

当然，在遍历集合元素时，也可同时指定 begin、end、step 3 个属性来确定迭代的范围。如示例 8-8 所示的代码片段。

示例 8-8：

```
…
<c:forEach var="accs" items="${accs}" begin="2" end="8" step="2">
   <tr>
     <td><c:out value="${accs.accId}"/></td>
<td><c:out value="${accs.accName}"/></td>
     <td><c:out value="${accs.accPwd}"/></td>
   </tr>
</c:forEach>
…
```

上面的代码指定了从索引为 2 的对象开始迭代输出，并且步长为 2，到索引为 8 的迭代结束。当然迭代的用法还不止这几种，大家应该先掌握这几种基本的使用方法，在日后再考虑扩展的使用方法。

<c:forTokens>标签专门用于处理字符串的迭代，可以指定一个或者多个分隔符。语法如下：

```
<c:forTokens items="items" delims="delimiters" [var="varName"]
  [varStatus="varStatusName"] [begin="begin"] [end="end"] [step="step"]>
  标签体内容
</c:forTokens>
```

其中参数的解释如表 8-6 所示。

表 8-6

参　数	作　用
var	指定迭代的参数的名字
items	指定将要迭代的 String
varStatus	表示迭代的状态，可以访问迭代自身的信息
delims	分隔符

下面给出示例 8-9 所示的例子，要分隔的字符串是“bule,red,green|yellow|pink,black|white”，分别使用“|”“|,”“_”作为分隔符。

示例 8-9：

```
<%@ page language="java" contentType="text/html; charset=UTF-8"
    pageEncoding="UTF-8"%>
    <%@taglib prefix="c" uri="http://java.sun.com/jsp/jstl/core" %>
<!DOCTYPE html>
<html>
<head>
<meta charset="UTF-8">
<title>Insert title here</title>
</head>
<body>
  <h4>使用"|"作为分隔符</h4>
  <c:forTokens var="token" items="bule,red,green|yellow|pink,black|white" delims="|">
  <c:out value="${token}"/>&copy;
  </c:forTokens>
<h4>使用"|,"作为分隔符</h4>
  <c:forTokens var="token" items="bule,red,green|yellow|pink,black|white" delims="|,">
  <c:out value="${token}"/>&copy;
  </c:forTokens>
  <h4>使用"_"作为分隔符</h4>
  <c:forTokens var="token" items="bule,red,green|yellow|pink,black|white" delims="_">
  <c:out value="${token}"/>&copy;
  </c:forTokens>
</body>
</html>
```

运行结果如图 8-8 所示。

图 8-8

要注意的是，可以使用一个字符作为分隔符，如使用“|”作为分隔符；也可以使用多个字符作为分隔符，如使用“|”和“,”作为分隔符。

8.5　国际化和格式化标签库

国际化的英文是 Internationalization，在第一字母 I 和最后一个字母 N 之间共有 18 个字母，所以简称 I18N。国际化是指在软件的设计阶段，就应该使软件具有支持多种语言和国家或地区的功能。这样，当需要在软件中添加对新的语言和国家的支持时，不需要重新修改原来的软件代码。一个软件支持国际化，应该具备下述特征：

- 当软件需要支持新的语言时，不需要修改其代码。
- 文本、消息和图片从源程序中提取出来，存储在外部。
- 软件应该根据用户的语言和地理位置，对与特定文化相关的数据，如日期、时间和货币，进行格式化。
- 支持非标准的字符集。

Web 应用程序中的国际化是指 Web 服务器可以根据不同客户端当地的语言环境来显示与语言环境匹配的页面内容和对数字、货币等进行格式化，符合当地用户的文化习惯。使用过 Google 搜索引擎的读者可能会有这样的体验，当客户端的语言环境设置成英文时，此时，访问 Google 站点，则页面全部是英文的内容。如果把语言环境改回到中文环境，则刷新页面，此时页面是中文内容。这样做的目的是使站点不仅可以面向国内的用户群，同时也可以面向国外用户群，所以，称之为国际化 I18N。

大多数 Web 浏览器允许用户配置 Locale，例如，可以通过更改 IE 浏览器的语言设置来修改客户端的语言环境，如图 8-9 所示。

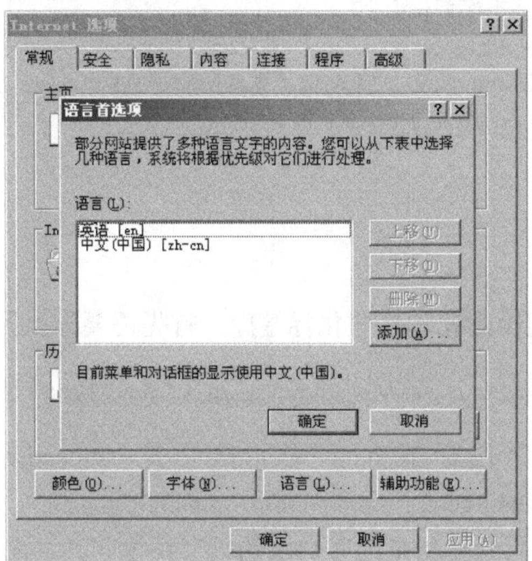

图 8-9

8.5.1　支持国际化的 Java 类

Web 应用程序的国际化依赖于两个特定的 Java 类，分别是 java.util.Locale 类和 java.util.

ResourceBundle 类。

Locale 类的实例表示了特定的地理、政治和文化地区，如果 Java 程序中的某个类在运行时需要根据 Locale 对象来调整其功能，那么，我们称这个类是本地敏感的(Locale Sensitive)。例如，我们后面将要提到的 java.text.DateFormat 类就是本地敏感的，因为它需要根据特定的 Locale 对象来对日期进行相应的格式化。Locale 对象本身并不执行和国际化相关的格式化或解析工作，它仅仅负责向本地敏感的类提供本地化信息，因为 Java 中使用 Locale 对象来存储与特定语言环境相关的信息。

ResourceBundle 类称为资源包，它包含特定于语言环境的对象。当程序需要一个特定于语言环境的资源时(如 String)，程序可以从适合当前用户语言环境的资源包中加载它。使用这种方式，可以编写在很大程度上独立于用户语言环境的程序代码，它将资源包中大部分(即便不是全部)特定于语言环境的信息隔离开来。使用资源包的好处是，开发者可以将应用程序轻松地本地化或翻译成不同的语言，一次处理多个语言环境，方便以后可以轻松修改，以便支持更多的语言环境。

Java 虚拟机在启动时会自动查询本地的操作系统，为运行环境设置默认的 Locale。Web 容器在其本地环境中通常会使用默认的 Locale，对于特定的客户端，Web 容器会从 HTTP 请求头中获取 Locale 信息。所以在 Web 应用中，用户不需要创建 Locale 对象实例，Web 容器会负责创建 Locale 实例。那么在 JSP 页面中，要获取客户端浏览器的首选的语言环境或支持多语言环境对象集合，可以使用下面的代码，其中两个方法都会访问 HTTP 请求头中的 Accept-Language 头信息。

```
//返回客户浏览器的首选的语言环境
Locale local = request.getLocale();
//返回支持多语言环境 Enumeration 对象，按优先级降序排列的 Locale 对象集合
Enumeration enum = request.getLocales();
```

8.5.2 国际化标签

要在 JSP 页面中使用国际化和格式化标签库，首先需要导入国际化和格式化的 URI。

```
<%@ taglib uri= " http://java.sun.com/jsp/jstl/fmt " prefix= " fmt " %>
```

JSTL 中的国际化常用标签有 4 个，分别是<fmt:setLocale>、<fmt:bundle>、<fmt:setBoundle>和<fmt:message>。

< fmt:setLocale>标签用于重新指定客户端的语言环境。其语法为：

```
<fmt:setLocale value="value"
[scope="page|request|session|application"] [variant=" variant "]/>
```

其中参数的解释如表 8-7 所示。

表 8-7

参 数	作 用
value	Locale 环境的指定，可以是 java.util.Locale 或 String 类型的实例
scope	指定变量的存储作用域，可选
variant	指定特定于浏览器的变量，可选

value 参数如果是 String 类型的，通常包含一个含有两个小写字母的语言代码和一个含有两个大写字母的国家或地区代码，语言代码和国家或地区代码之间用下画线连接。在 IE 中常用的语言和国家或地区代码如表 8-8 所示。

表 8-8

代 码	国家或地区	代 码	国家或地区
ar_kw	阿拉伯语(科威特)	de	德语(德国)
ar_iq	阿拉伯语(伊拉克)	de_ch	德语(瑞士)
fr	法语(法国)	en	英语(英国)
fr_ch	法语(瑞士)	en_us	英语(美国)
es_pa	西班牙语(巴拿马)	zh	中文(中国大陆)
es_mx	西班牙语(墨西哥)	zh_w	中文(中国台湾)

语言代码遵循 ISO-639 规范，可以从 http://www.unicode.org/onlinedat/languages.html 获取完整的语言代码列表。国家代码遵循 ISO-3166 规范，可以从 http://www.unicode.org/onlinedat/ countries.html 获取完整的国家代码列表。

例如，<fmt:setLocale value="zh_tw"/>表示设置客户端的本地环境为繁体中文。

<fmt:bundle>标签用来创建 I18N 的本地化上下文，并将资源包加载在其中，资源包的名称由<fmt:bundle>标签中的 basename 属性指定。其语法为：

```
<fmt:bundle basename="basename">
标签体内容
</fmt:bundle>
```

<fmt:setBundle>标签用来创建 I18N 的本地化上下文，并将资源包存储在某个作用域范围中，其语法为：

```
<fmt:setBoundle basename="basename"
[scope="page|request|session|application"] var="varName "/>
```

其中的 basename 是指定资源包的名称，var 指定导出的变量的名称，它存储在 I18N 本地上下文中，scope 指定 var 的范围。

这两个标签都用于资源配置文件的绑定，唯一不同的是<fmt:bundle>标签将资源配置文件绑定于它的标签体中的显示，<fmt:setBundle>标签则允许将资源配置文件保存为一个变量，之后的工作可以根据该变量来进行。根据不同的 Locale 环境来查找不同后缀的资源配置文件，这在国际化的任何技术上都是一致的。通常来说，这两种标签单独使用是没有意义的，它们都会与 I18N 格式化标签库中的其他标签配合使用。

例如下面的代码片段：

```
<fmt:setLocale value="zh_CN"/>
<fmt:setBundle basename="applicationMessage"
var="applicationBundle"/>
```

它的作用将会查找一个名为 applicationMessage_zh_CN.properties 的资源配置文件，来作为显示的 Resource 绑定。

<fmt:message>标签用来给出资源包的输出值。其语法为：

```
<fmt:message key="key" bundle="资源包"
[scope="page|request|session|application"] [var ="varName"] />
```

其中参数的解释如表 8-9 所示。

表 8-9

参　数	作　用
key	资源包中的键指定的值
bundle	若使用<fmt:setBundle>标签指定了资源配置文件，该属性就可以从该资源配置文件中进行查找
var	将显示信息保存到一个变量
scope	变量的作用范围

来看下面的代码片段：

```
<fmt:setBundle basename="applicationMessage"
var="applicationBundle"/>
<fmt:bundle basename="applicationAllMessage">
    <fmt:message key="userName" />
    <br>
    <fmt:message key="passWord" bundle="${applicationBundle}" />
</fmt:bundle>
```

示例中使用了两种资源配置文件的绑定的做法，“applicationMessage”资源配置文件使用<fmt:setBundle>标签被赋予了变量“applicationBundle”，而作为<fmt:bundle>标签定义的“applicationAllMessage”资源配置文件作用于其标签体内的显示。第一个<fmt:message>标签将使用“applicationAllMessage”资源配置文件中“键”为“userName”的信息显示。第二个<fmt:message>标签虽然被定义在<fmt:bundle>标签体内，但是它使用了 bundle 属性，因此将使用由<fmt:setBundle>标签指定的“applicationMessage”资源文件，而不是使用“applicationAllMessage”资源文件，并在页面上显示键为“passWord”的信息。

在 Web 应用中使用国际化，通常需要以下步骤。

(1) 建立默认的资源文件，内容是键值对的形式。

(2) 根据要使用的国家或地区建立与之对应的资源文件，注意资源文件的名字必须是默认的资源文件名+“_”+“语言和国家或地区代码”。

(3) 在页面中使用国际化标签。

这里要特别强调，资源文件的内容不支持中文，需要把中文内容转码成 UTF-8 的编码。使用 JDK 里的 native2ascii 命令，在 cmd 的命令提示符下，输入“native2ascii –encoding gb2312 xxx.properties xxx_zh_cn.properties”即可把资源文件转成 UTF-8 编码。在 MyEclipse 开发环境中不需要，它会自动帮用户转换。

下面来看一个账号注册页面的国际化例子。默认的资源文件名为 message.properties，放在 Web 工程的 src 目录下面，另外再创建一个名称为 message_en.properties 的文件，内容与 message.properties 一样，如示例 8-10 所示。

示例 8-10：

```
page_title0=Bank Account register page
page_mes1=Please input your personal information:
page_mes2=Account Id:
page_mes3=Your name:
page_mes4=Your password:
page_mes5=Initial Balance:
page_action1=submit
```

新建 message_zh_CN.properties 资源文件，同样存放在 src 目录下面，此时，转到 Design 视图，添加和示例 8-10 所示的键值对，转到 Source 视图，则输入的中文被自动转换成 UTF-8 的编码，如示例 8-11 所示。

示例 8-11：

```
page_title0=\u94F6\u884C\u5E10\u53F7\u6CE8\u518C\u9875\u9762
page_mes1=\u8BF7\u8F93\u5165\u4F60\u7684\u4E2A\u4EBA\u4FE1\u606F
page_mes2=\u5E10\u53F7\uFF1A
page_mes3=\u540D\u5B57\uFF1A
page_mes4=\u5BC6\u7801\uFF1A
page_mes5=\u5F00\u6237\u91D1\u989D\uFF1A
page_action1=\u6CE8\u518C
```

使用国际化标签的 JSP 注册页面如示例 8-12 所示。

示例 8-12：

```
<%@ page contentType="text/html; charset=gbk"%>
<%@ taglib prefix="c" uri="http://java.sun.com/jsp/jstl/core"%>
<%@ taglib prefix="fmt" uri="http://java.sun.com/jsp/jstl/fmt"%>
<html>
```

```
    <head>
            <!--指定页面使用的资源文件的名称-->
            <fmt:setBundle basename="message"/>
        <title><fmt:message key="page_title0" /></title>
    </head>
    <body>
    <form action="showinfo.jsp" method="post">
        <fmt:message key="page_mes1"/> <br>
        <fmt:message key="page_mes2"/><input type="text" name="accId"><br>
        <fmt:message key="page_mes3"/><input type="text" name="accName"><br>
        <fmt:message key="page_mes4"/><input type="password" name="accPwd"><br>
        <fmt:message key="page_mes5"/><input type="text" name="balance"><br>
        <input type="submit" value='<fmt:message key="page_action1"/>'>
   </form>
   </body>
</html>
```

部署该 Web 应用到 Tomcat 服务器并启动，同时设置 IE 浏览器的语言为“英语[en]”，页面的运行结果如图 8-10 所示。

这时，如果更改 IE 浏览器的设置，把语言改成“中文(中国)[zh_cn]”，单击“刷新”按钮，则页面的显示由英文内容变成了中文内容，如图 8-11 所示。

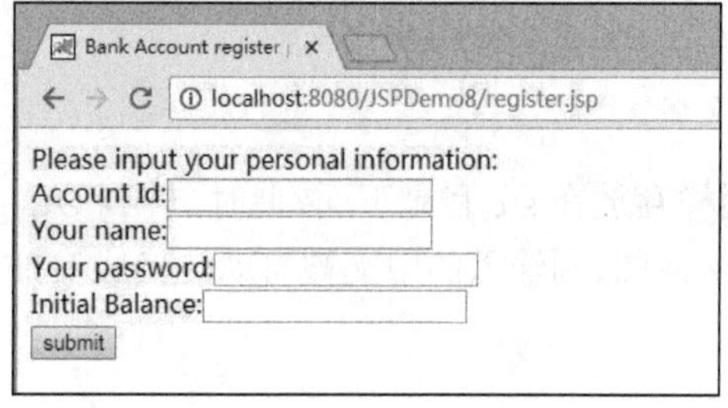

图 8-10

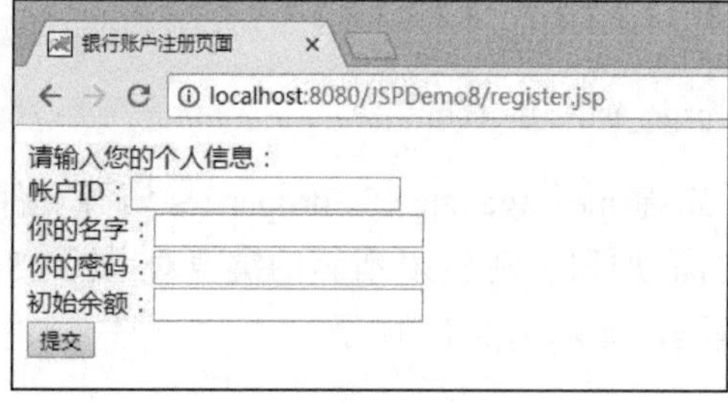

图 8-11

8.5.3 支持格式化的 Java 类

Web 应用程序的格式化标签依赖于 3 个特定的 Java 类，分别是 java.util.TimeZone、java.text.DateFormat 和 java.text.NumberFormat。

TimeZone 类的实例包含了一个与格林威治标准时间(GMT)相比较得出的以微秒为单位的时区偏移量，它还可以用来处理夏令时。在 Java 中，使用时区 ID 来表示一个时区。例如美国太平洋时区的时区 ID 是 America/Los_Angeles。因此，可以使用下面语句获得美国太平洋时间 TimeZone 对象实例：TimeZone tz = TimeZone.GetTimeZone("America/Los_Angeles")。当然，用户可以自己创建一个时区 ID，指定的自定义时区 ID 采用“GMT Sign TwoDigitHours : Minutes”的形式语法进行标准化，其中 Sign 的取值是“+”或“-”，分别表示比 GMT 的时间早还是晚。TwoDigitHours 和 Minutes 两个参数的取值是 0、1、2、3、4、5、6、7、8、9 这 10 个数字。例如：TimeZone.getTimeZone("GMT+8")

得到的是北京、重庆和香港的时区对象的实例。

DateFormat 类是用来格式化日期/时间的抽象类，它以与语言无关的方式格式化并解析日期或时间。在 Java 中，我们将日期表示为 Date 对象，或者表示为从 GMT(格林尼治标准时间)1970 年 1 月 1 日 00:00:00 这一刻开始的毫秒数。使用 DateFormat 提供的类方法，可以获得基于默认或给定语言环境和多种格式化风格的默认日期/时间 Formatter，用来格式化 Date 对象。格式化风格包括 FULL、LONG、MEDIUM 和 SHORT，其中，SHORT 完全为数字，如 12.13.52 或 3:30pm；MEDIUM 较长，如 Jan 12, 1952；LONG 更长，如 January 12, 1952 或 3:30:32pm；FULL 是完全指定，如 Tuesday、April 12、1952 AD 或 3:30:42pm PST。

DateFormat 可帮助进行格式化并解析任何语言环境的日期，在实际的应用中，我们一般使用得比较多的是 DateFormat 类的子类 SimpleDateFormat，因为它能够很方便地使用自定义日期和时间的模式来格式化 Date 对象。常用的日期和时间模式如表 8-10 所示。

表 8-10

日期和时间模式	结　果
"yyyy.MM.dd G 'at' HH:mm:ss z"	2001.07.04 AD at 12:08:56 PDT
"EEE, MMM d, ''yy"	Wed, Jul 4, '01
"h:mm a"	12:08 PM
"hh 'o''clock' a, zzzz"	12 o'clock PM, Pacific Daylight Time
"K:mm a, z"	0:08 PM, PDT
"yyyyy.MMMMM.dd GGG hh:mm aaa"	02001.July.04 AD 12:08 PM
"EEE, d MMM yyyy HH:mm:ss Z"	Wed, 4 Jul 2001 12:08:56 -0700
"yyMMddHHmmssZ"	010704120856-0700
"yyyy-MM-dd'T'HH:mm:ss.SSSZ"	2001-07-04T12:08:56.235-0700

NumberFormat 类是所有数值格式的抽象基类，该类提供格式化和解析数值的接口。NumberFormat 类可用于格式化和解析任何语言环境的数值，使代码能够完全独立于小数点、千位分隔符甚至所用特定小数位数的语言环境约定，并与数值格式是否为偶小数无关。在实际的应用中，我们使用比较多的是 NumberFormat 类的子类 DecimalFormat 类，因为我们可以很容易地使用自定义的数字模式来格式化数字。DecimalFormat 类中定义的模式符号如表 8-11 所示。

表 8-11

数字格式模式符号	含　义
0	表示一个数位
#	表示一个数位，前导零和追尾零不显示
.	表示小数点分割位置
%	表示用 100 乘，并显示百分号
,	表示组分隔符的位置
-	表示负数前缀

8.5.4 格式化标签

JSTL 中的格式化常用标签有 6 个，分别是<fmt:setTimeZone>、<fmt:timeZone>、<fmt:formatDate>、<fmt:parseDate>、<fmt:formatNumber>和<fmt:parseNumber>。

(1) <fmt:setTimeZone>标签用于设定页面时间的时区。其语法为：

```
<fmt:setTimeZone value="value" [scope="page|request|session|application"] var=" varName "]/>
```

其中参数的解释如表 8-12 所示。

表 8-12

参　数	作　用
value	设置时区的值，可以是 java.util.TimeZone 的实例或 String 类型的时区 ID
var	把当前的时区保存到指定的变量中
scope	指定变量的存储作用域

(2) <fmt:timeZone>标签用于在标签体内使用 value 指定的时区。其语法为：

```
<fmt:timeZone value="value" >
标签体内容
</fmt:timeZone>
```

以上两个标签都用来指定时区，区别是，使用<fmt:setTimeZone>标签来指定时区，作用域是对于该标签以下的页面内容，而<fmt:timeZone>标签的作用域仅限于其标签体的内容。

(3) <fmt:formatDate>标签用于格式化日期，其语法为：

```
<fmt:formatDate type="type" var="varName" timeZone="timeZone" pattern="pattern"
scope="page|request|session|application" value="value" dateStyle="style" timeStyle
="style" />
```

其中参数的解释如表 8-13 所示。

表 8-13

参　数	作　用
value	格式化的日期，该属性的内容是 java.util.Date 类的实例
type	格式化的类型，取值是 date、time 或 both
pattern	格式化模式，参考 java.text.SimpleDateFormat 类的定义的模式
var	结果保存变量，类型为 java.util.Date 类型
scope	变量的作用范围
timeZone	指定格式化日期的时区
dateStyle	输出日期的类型，取值是 default、short、medium、long 或 full
timeStyle	输出时间的类型，取值是 default、short、medium、long 或 full

示例 8-13 所示的代码演示了<fmt:timeZone>和<fmt:formatDate>标签的用法。

示例 8-13：

```
<%@ page language="java" contentType="text/html; charset=UTF-8"
    pageEncoding="UTF-8"%>
    <%@taglib prefix="fmt" uri="http://java.sun.com/jsp/jstl/fmt" %>
<!DOCTYPE html>
<html>
<head>
<meta charset="UTF-8">
<title>设置时区和格式化日期、时间</title>
</head>
<body>
 <fmt:setLocale value="zh_CN"/>
      <fmt:timeZone value="GMT+8">
       <jsp:useBean id="now" class="java.util.Date" scope="page" />
       <fmt:formatDate value="${now}" type="date" /><br>
       <fmt:formatDate value="${now}" type="time" /><br>
       <fmt:formatDate value="${now}" type="both" /><br>
       <fmt:formatDate value="${now}" type="date" timeStyle="default" /><br>
       <fmt:formatDate value="${now}" type="date" dateStyle="full" /><br>
    <fmt:formatDate value="${now}" type="time" timeStyle="default" /><br>
       <fmt:formatDate value="${now}" type="time" timeStyle="short" /><br>
       <fmt:formatDate value="${now}" type="time" timeStyle="medium" /><br>
       <fmt:formatDate value="${now}" type="time" timeStyle="long" /><br>
       <fmt:formatDate value="${now}" type="time" timeStyle="full" /><br>
       <fmt:formatDate value="${now}" type="both"
          pattern="EEEE, MMMM d, yyyy HH:mm:ss Z" /><br>
       <fmt:formatDate value="${now}" type="both"
          pattern="d MMM yy, h:m:s a zzzz" /><br>
       </fmt:timeZone>

</body>
</html>
```

运行结果如图 8-12 所示。

示例 8-13 的代码时区是“GMT+8”，北京时区，语言环境是中文。如果把时区更改成“GMT-8”，语言环境改成“en_US”，则运行结果如图 8-13 所示。

图 8-12

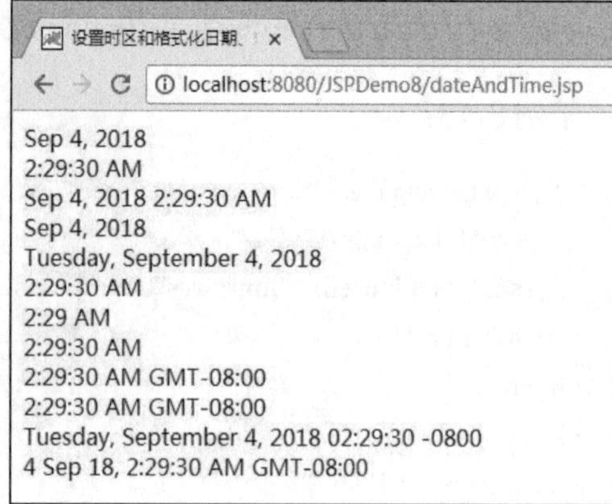

图 8-13

(4) <fmt:parseDate>标签用于解析一个日期，并将结果作为 java.lang.Date 类型的实例返回。<fmt:parseDate>标签和<fmt:formatDate>标签的作用正好相反。其语法为：

```
<fmt:parseDate dateStyle="style" parseLocale="locale" pattern="pattern" scope="
page|request|session|application" timeStyle="style" timeZone="timezone"    type="type"
value="value" var="varName">
标签体内容
</fmt:parseDate>
```

其中参数的解释如表 8-14 所示。

表 8-14

参　数	作　用
value	将被解析的字符串
type	格式化的类型
pattern	格式化模式，参考 java.text.SimpleDateFormat 类的定义的模式
var	结果保存变量，类型为 java.util.Date 类型
scope	变量的作用范围
timeZone	指定格式化日期的时区
parseLocale	以本地化的形式来解析字符串，该属性的内容为 String 或 java.util.Locale 类的实例

(5) <fmt:formatNumber>标签用来格式化数字，实际上是对应 java.util.NumberFormat 类，其语法为：

```
<fmt:formatNumber pattern="pattern" type="type" value="expression" maxFractionDigits=""
maxIntegerDigits=""    minFractionDigits="" minIntegerDigits=""    scope="page|request|session|application "
var="varName">
标签体内容
</fmt:formatNumber>
```

其中参数的解释如表 8-15 所示。

表 8-15

参　数	作　用
value	指定要格式化的数字，该数值可以是 String 类型或 java.lang.Number 类型的实例
type	格式化的类型，值包括 currency(货币)、number(数字)和 percent(百分比)
pattern	格式化模式
var	结果要保存变量
scope	变量的作用范围
maxIntegerDigits	指定格式化结果的最大值
minIntegerDigits	指定格式化结果的最小值
maxFractionDigits	指定格式化结果的最大值，带小数
minFractionDigits	指定格式化结果的最小值，带小数

我们来看下面的代码片段和输出。

```
<fmt:formatNumber value="12" type="currency" pattern=".00 元"/>
 将显示 12.00 元
<fmt:formatNumber value="12" type="currency" pattern=".0#元"/>
 将显示 12.00 元
<fmt:formatNumber value="1234567890" type="currency"/>
 将显示￥1,234,567,890.00  (当前 Web 服务器的语言环境设定为 zh_CN)
<fmt:formatNumber value="123456.7891" pattern="#,#00.0#"/>
 将显示 123,456.79
<fmt:formatNumber value="12" type="percent" />
 将显示 1,200%
```

(6) <fmt:parseNumber>标签用于解析一个数字，并将结果作为 java.lang.Number 类的实例返回。<fmt:parseNumber>标签和<fmt:formatNumber>标签的作用正好相反。其语法为：

```
<fmt:parseNumber pattern="pattern" scope=" page|request|session|application" type="type" value="value"
    var="varName" parseLocale ="locale">
标签体内容
</fmt:parseNumber>
```

其中参数的解释如表 8-16 所示。

表 8-16

参　数	作　用
value	将要解析的字符串
type	格式化的类型
pattern	格式化模式

(续表)

参　数	作　用
var	结果保存变量，类型为 java.lang.Number
scope	变量的作用范围
parseLocale	以本地化的形式来解析字符串，该属性的内容应为 String 或 java.util.Locale 类型的实例

【单元小结】

- 通用标签用于在 JSP 页面中创建、删除和显示其中的变量值。
- 条件标签用于支持 JSP 页面中的各种条件。
- 迭代标签用于遍历某个集合中的元素。
- 国际化标签用来根据客户端的不同语言环境显示不同的页面内容。
- 格式化标签可以格式化数字、时间和货币。

【单元自测】

1. (　　)用于操作 XML 文档，访问数据库，以及执行其他功能。

   A. 自定义标签　　B. Scriptlet　　C. 指令　　D. JSTL

2. (　　)用来在 JSP 页面中创建、删除和显示变量值。

   A. 条件标签　　B. 迭代标签　　C. 常用标签　　D. 自定义标签

3. (　　)支持 JSP 页面中的条件。

   A. 自定义标签　　B. 常用标签　　C. 条件标签　　D. 自定义标签

4. (　　)多次计算它的标签体。

   A. 迭代标签　　B. 标准操作标签

   C. 常用标签　　D. 条件标签

5. <c:forTokens>标签的作用是(　　)。

   A. 重复完成一段代码　　B. 判断条件是否满足

   C. 将字符串按预定义分割　　D. 设置变量的值

【上机实战】

上机目标

- 掌握国际化和格式化标签的使用。
- 掌握核心标签库中的标签的使用。

上机练习

◆ 第一阶段 ◆

练习 1：仿 Google 首页的国际化页面

【问题描述】

编写一个仿 Google 首页的页面，页面上有两个超链接，分别是“Google.com in English”和“Go to Google China”。当单击“Google.com in English”时，首页的内容全部变成英文；当单击“Go to Google China”时，页面的内容变成中文。

【问题分析】

这是一个典型的国际化页面，需要用到资源文件来存放页面的内容，单击不同的超链接，实际上是改变页面作用域内的语言设置，同时把页面提交给自己，然后根据不同的语言环境去读取资源文件中的信息，显示在页面上。

【参考步骤】

(1) 创建默认的资源文件 information.properties 和 information_en.properties，如示例 8-14所示，注意两个文件都存放在工程的 src 目录下面。

示例 8-14：

```
info0=Web
info1=Images
info2=Maps
info3=News
info4=Shopping
info5=GMail
info6=more
btn1=Google Search
btn2=I am Feeling Luckly
info7=iGoogle
info8=Sign in
info9=Advanced Search
info10=Preferences
info11=Language Tools
info12=Advertising Programs
info13=Business Solutions
info14=About Google
info15=Go to Google China
```

(2) 创建中文环境下的资源文件 information_zh_CN.properties，如示例 8-15 所示。

示例 8-15：

```
info0=\u7F51\u9875
info1=\u56FE\u7247
info2=\u5730\u56FE
info3=\u65B0\u95FB
info4=\u89C6\u9891
info5=\u535A\u5BA2
info6=\u66F4\u591A
btn1=Google \u641C\u7D22
btn2=\u624B\u6C14\u4E0D\u9519
info7=\u4E2A\u6027\u5316\u9996\u9875
info8=\u767B\u5F55
info9=\u9AD8\u7EA7\u641C\u7D22
info10=\u4F7F\u7528\u504F\u597D
info11=\u8BED\u8A00\u5DE5\u5177
info12=\u5E7F\u544A\u8BA1\u5212
info13=\u751F\u610F\u4E4B\u9053
info14=\u5173\u4E8EGoogle
info15=Google in English
```

(3) 创建 google.jsp，代码如示例 8-16 所示。

示例 8-16：

```
<%@ page language="java" contentType="text/html; charset=UTF-8"
    pageEncoding="UTF-8"%>
    <%@ taglib prefix="c" uri="http://java.sun.com/jsp/jstl/core"%>
<%@ taglib prefix="fmt" uri="http://java.sun.com/jsp/jstl/fmt"%>
<!DOCTYPE html>
<html>
  <head> <title>Google</title>
    <style type="text/css">
    a{ font-family:宋体; font-size:13px;}
    </style>
    <script type="text/javascript">
     function fun()
     {
        var value = document.getElementById("a1").innerText;
        if(value=="Go to Google China"){
        document.getElementById("a1").href="google.jsp?lan=zh_CN";
        }else{
          document.getElementById("a1").href="google.jsp?lan=en";
        }
     }
    </script>
</head>
<body>
  <c:if test="${!empty param.lan }">
    <fmt:setLocale value="${param.lan}"/>
```

```
    </c:if>
    <fmt:setBundle basename="information"/>
     <font size="1" ><fmt:message key="info0"/></font>
     <a href="#"><fmt:message key="info1"/></a>
     <a href="#"><fmt:message key="info2"/></a>
     <a href="#"><fmt:message key="info3"/></a>
     <a href="#"><fmt:message key="info4"/></a>
     <a href="#"><fmt:message key="info5"/></a>
     <a href="#"><fmt:message key="info6"/></a>
           
     <a href="#"><fmt:message key="info7"/></a>
     <a href="#"><fmt:message key="info8"/></a>
     <hr>
     <form action="" method="post" style="overflow: auto;">
     <img src="logo.jpg" style="display: inherit;"/>
     <div style="float:left;">
     <input type="text"name="input" size ="50"><br>
     <input type="button" value='<fmt:message key="btn1"/>'>
     <input type="button" value='<fmt:message key="btn2"/>'>
     </div>
     <div style="float:left;">
     <a href="#"><fmt:message key="info9"/></a><br>
     <a href="#"><fmt:message key="info10"/></a><br>
     <a href="#"><fmt:message key="info11"/></a><br>
     </div>
     <br><br>
     </form>
     <a href="#"><fmt:message key="info12"/></a>
     <a href="#"><fmt:message key="info13"/></a>
     <a href="#"><fmt:message key="info14"/></a>
     <a id ="a1" href="#" onclick="fun()"><fmt:message key="info15"/></a>
  </body>
  </html>
```

(4) 页面的运行结果如图 8-14 所示。

图 8-14

(5) 此时，单击页面上的“Google in English”超链接，页面的内容转成英文，如图 8-15 所示，此时，如果单击页面上的“Go to Google China”超链接，页面的内容将显示成中文。

图 8-15

练习 2：使用核心标签库

【问题描述】

创建一个 JSP 页面，此页面将创建一个 ArrayList 集合对象，在此集合中存储 JavaBean 信息，并使用核心标签库的标签以表格的形式显示集合中的内容。

【问题分析】

我们经常会碰到类似的问题，Servlet 控制器从数据库取出表中的数据，并把数据存储在某个作用域里面，然后把页面转向到显示数据的页面，在页面里显示这些数据。使用 JSTL 的<c:forEach>标签和<c:out>标签迭代输出集合中的每个 JavaBean 对象。

【参考步骤】

创建 showBeanInfo.jsp 页面，代码如示例 8-17 所示。

示例 8-17：

```
<%@page contentType="text/html; charset=gbk"%>
<%@page import="java.util.*,bean.UserBean"%>
<%@taglib prefix="c" uri="http://java.sun.com/jstl/core"%>
<%
  List users = new ArrayList();
  for (int i = 0; i < 10; i++) {
    UserBean user = new UserBean();
    user.setUserName("guest" + i);
    user.setUserPwd("guest1" + i);
    user.setAge(20 + i);
    users.add(user);
  }
```

```
    session.setAttribute("users", users);
%>
<html>
<head>
<title>c:forEach 标签和 c:out 标签</title>
</head>
<body bgcolor="#FFFFFF">
<center>
  <h4>User list</h4>
  <table border=1>
    <tr>
      <td>用户名</td>
      <td>密码</td>
      <td>年龄</td>
    </tr>
    <c:forEach var="users" items="${users}">
      <tr>
        <td>
          <c:out value="${users.userName}"/>
        </td>
        <td>
          <c:out value="${users.password}"/>
        </td>
        <td>
          <c:out value="${users.age}"/>
        </td>
      </tr>
    </c:forEach>
  </table>
</center>
</body>
</html>
```

◆ 第二阶段 ◆

练习 3：使用格式化标签格式化时间和货币

【问题描述】

编写一个页面，用户按照格式输入时间和货币，然后选择地点(美国、德国、俄罗斯和法国)，在第二个页面输出与地点对应的日期时间和货币的格式。

【问题分析】

(1) 第一个页面是一个普通的含有表单的页面，用来接收用户按照格式输入的时间和数字。

(2) 根据用户选择的地点，来设置时区和 Locale，在页面上输出格式化后的时间和货币。

【拓展作业】

1. 使用<c:forTokens>标签输出客户端的 IP 地址中的每个数字。
2. 使用<c:forEach>标签输出客户端浏览器所支持的语言环境。

课程目标

- 理解 Servlet 过滤器
- 理解 Servlet 过滤器的生命周期
- 掌握实现简单的 Servlet 过滤器的方法

简 介

Servlet 过滤器是在 Servlet 规范 2.3 中定义的，它能够对 Servlet 容器的请求和响应对象进行检查和修改。Servlet 过滤器本身并不产生请求和响应对象，它只是提供过滤作用。Servlet 过滤器能够在 Servlet 被调用之前检查 Request 对象，修改 Request Header 和 Request 内容；在 Servlet 被调用之后检查 Response 对象，修改 Response Header 和 Response 内容。

9.1 Servlet 过滤器的机制和特点

Servlet 过滤器负责过滤的 Web 组件有 Servlet、JSP 或者是 HTML 文件。其过滤过程如图 9-1 所示。

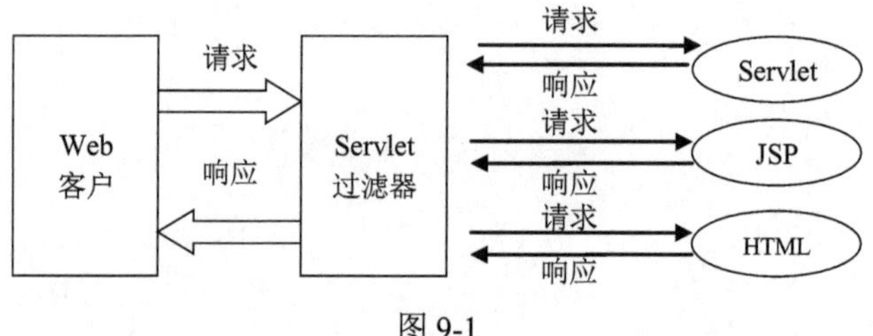

图 9-1

Servlet 过滤器具有以下特点。

- Servlet 过滤器可以检查和修改 ServletRequest 和 ServletResponse 对象。
- 可以指定 Servlet 过滤器和特定的 URL 关联，只有当客户请求访问此 URL 时，才会触发过滤器工作。
- Servlet 过滤器是 Servlet 规范 2.3 的一部分，因此所有实现 Servlet 规范 2.3 的 Servlet 容器都支持 Servlet 过滤器。
- 独立 Servlet 过滤器可以被串在一起，形成管道效应，协同修改请求和响应对象。

9.2 过滤器的生命周期

Servlet 过滤器的生命周期与前面讲的 Servlet 一样，包含了如下几个阶段。

- 实例化：在访问 Web 资源之前创建过滤器的实例。
- 初始化：Web 容器在调用过滤器的 doFilter()方法之前调用 init()方法，把 FilterConfig 对象作为入参传给 init()方法。
- 过滤：每当用户提交请求或 Web 资源发送响应时，调用 doFilter()方法。
- 销毁：在停止使用过滤器之前，由容器调用过滤器的 destroy()方法，允许过滤器完成必要的清除和释放资源。

9.3 过滤器的 API

所有的 Servlet 过滤器都必须实现 javax.servlet.Filter 接口。在这个接口里定义了 3 个过滤器类必须实现的方法。

(1) init()：Servlet 过滤器的初始化方法。Servlet 容器创建 Servlet 过滤器实例后将调用这个方法。在这个方法里可以读取 web.xml 文件中 Servlet 过滤器的初始化参数。init()方法的语法为：

```
public void init(FilterConfig filterConfig) throw ServletException{}
```

(2) doFilter()：这个方法完成实际的过滤器操作。当客户请求访问与过滤器关联的 URL 时，Servlet 容器将先调用过滤器的 doFilter()方法。FilterChain 参数用于访问后续过滤器。doFilter()方法的语法为：

```
public void doFilter (ServletRequest request ,ServletResponse response ,FilterChian chain) throws
IOEcxeption ,ServletException{}
```

(3) destroy()：Servlet 容器在销毁过滤器前调用该方法，这个方法中可以释放 Servlet 过滤器占用的资源。destroy()方法的语法为：

```
public void destroy (){}
```

9.4 实现过滤器

下面举一个例子，同时来看看如何在 Eclipse 环境中开发 Filter。本例中的 Filter 叫 BlacklistFilter，可以实现拒绝列在黑名单上的客户访问 Web 资源，同时能将服务器响应客户的请求所花的时间写入日志。

9.4.1 创建 Servlet 过滤器

MyEclipse 中没有相应的 Filter 选项直接创建 Filter，因此需要自己来写。写一个类继承 javax.servlet.HttpServlet 和实现 javx.servlet.Filter 接口。代码如示例 9-1 所示。

示例 9-1：

```
package control;

import java.io.IOException;
import java.io.PrintWriter;

import javax.servlet.Filter;
import javax.servlet.FilterChain;
```

```
import javax.servlet.FilterConfig;
import javax.servlet.ServletContext;
import javax.servlet.ServletException;
import javax.servlet.ServletRequest;
import javax.servlet.ServletResponse;
import javax.servlet.annotation.WebFilter;
import javax.servlet.http.HttpServlet;
import javax.servlet.http.HttpServletRequest;

@WebFilter("/BlacklistFilter")
public class BlacklistFilter extends HttpServlet implements Filter {
  private FilterConfig config = null;
  private String blackList = "";

  public void destroy() {
     this.config = null;
  }

  public void doFilter(ServletRequest request, ServletResponse response, FilterChain chain)
           throws IOException, ServletException {
     String username = request.getParameter("username");
     request.setCharacterEncoding("utf-8");
     response.setContentType("text/html;charset=utf-8");
     PrintWriter out = response.getWriter();
     if (username != null && username.indexOf(blackList) != -1) {
           out.println("<html><head><title>Filter 测试</title></head><body>");
           out.println("<h1>对不起, " + username + ",你没有权利使用 Web 资源</h1>");
           out.print("</body></html>");
     }
     // /计算 Web 服务器响应客户请求所花时间
     long begintime = System.currentTimeMillis();
     ServletContext context = this.config.getServletContext();
     context.log("BlacklistFilter 开始调用 doFilter()");
     chain.doFilter(request, response);
     context.log("BlacklistFilter 完成调用 doFilter()");
     long endtime = System.currentTimeMillis();
     String reqURI = "";
     if (request instanceof HttpServletRequest) {
           reqURI = ((HttpServletRequest) request).getRequestURI();
     }
     context.log("web 响应客户请求" + reqURI + "历时:" + (endtime - begintime) + "ms");
  }

  @Override
  public void init(FilterConfig config) throws ServletException {
     this.config = config;
     blackList = config.getInitParameter("blacklist");
```

```
    }
}
```

9.4.2　部署 Servlet 过滤器

部署 Servlet 过滤器时，必须要在 web.xml 文件中加入<filter>元素和<servlet-mapping>元素。<filter>元素用来定义一个过滤器。其中，<servlet-name>属性指定过滤器的名称，<filter-class>指定过滤器的类名，<init-param>子元素为过滤器提供初始化参数值，它包含一对参数名和参数值。在<filter>元素中可以包含多个<init-param>子元素。<filter>的语法为：

```
<filter>
    <description>过滤黑名单的 Filter</description>
    <filter-name>BlacklistFilter</filter-name>
    <filter-class>control.BlacklistFilter</filter-class>
    <init-param>
        <param-name>blacklist</param-name>
        <param-value>tom</param-value>
    </init-param>
</filter>
```

<filter-mapping>元素用于将该过滤器和 URL 关联，其中，<filter-name>必须要和上面<filter>中的<filter-name>一致，<url-pattern>指定匹配的 URL，代码中匹配 show.jsp。如果希望过滤器过滤所有的 URL，可以把<url-pattern>的值设成“/*”。<filter-mapping>的语法为：

```
<filter-mapping>
    <filter-name>BlacklistFilter</filter-name>
    <url-pattern>/show.jsp</url-pattern>
</filter-mapping>
```

完整的 web.xml 如示例 9-2 所示。

示例 9-2：

```
<?xml version="1.0" encoding="UTF-8"?>
<web-app xmlns:xsi="http://www.w3.org/2001/XMLSchema-instance"
xmlns="http://xmlns.jcp.org/xml/ns/javaee"
xsi:schemaLocation="http://xmlns.jcp.org/xml/ns/javaee
http://xmlns.jcp.org/xml/ns/javaee/web-app_3_1.xsd"
id="WebApp_ID" version="3.1">
<display-name>MyNinthWebApp</display-name>
<filter>
    <description>过滤黑名单的 Filter</description>
    <filter-name>BlacklistFilter</filter-name>
    <filter-class>control.BlacklistFilter</filter-class>
```

```
        <init-param>
            <param-name>blacklist</param-name>
            <param-value>tom</param-value>
        </init-param>
    </filter>
    <filter-mapping>
        <filter-name>BlacklistFilter</filter-name>
        <url-pattern>/show.jsp</url-pattern>
    </filter-mapping>

    </web-app>
```

9.4.3 测试 Servlet 过滤器

用两个 JSP 网页文件来测试过滤器，看看过滤器是否工作。第一个 JSP 页面中，要求用户输入用户名和密码，提交后，请求 show.jsp，在 show.jsp 页面中显示用户名，代码如示例 9-3 和示例 9-4 所示。

示例 9-3：

```
<%@ page language="java" contentType="text/html; charset=UTF-8"
    pageEncoding="UTF-8"%>
<!DOCTYPE html>
<html>
<head>
<meta charset="UTF-8">
<title>测试 Filter</title>
</head>
<body>
 <h1>欢迎登录</h1>
    <form action='show.jsp' method='POST'>
    用户名：<input type=text name=username value=''><br>
    密码：<input type=password name=userpwd value=''><br>
    <input type=submit name=submit value='提交'>
    </form>
    <br>
</body>
</html>
```

示例 9-4：

```
<%@ page language="java" contentType="text/html; charset=UTF-8"
    pageEncoding="UTF-8"%>
```

```
<!DOCTYPE html>
<html>
<head>
<meta charset="UTF-8">
<title>显示用户信息</title>
</head>
<body>
    <h1>欢迎你！<%=request.getParameter("username")%></h1> <br>
</body>
</html>
```

完成后在 IE 地址栏中输入“http://localhost:8080/MyNinthWebApp/login.jsp”，显示如图 9-2 所示的界面。

图 9-2

输入用户名 tom，密码可任意输入，显示如图 9-3 所示的界面，说明过滤器工作了，因为在 web.xml 中定义了黑名单 tom，如果用户是 tom，则不能访问网络资源。如果输入 tom 之外的用户名，例如 jack，则显示结果如图 9-4 所示。

图 9-3

图 9-4

9.5 使用 Servlet 过滤器链

可以把多个 Servlet 过滤器串联起来一起协同工作，Servlet 容器可以根据它们在 web.xml 中定义的先后顺序，依次调用它们的 doFilter()方法。下面来看看两个 Servlet 过滤器是如何工作的，流程如图 9-5 所示。

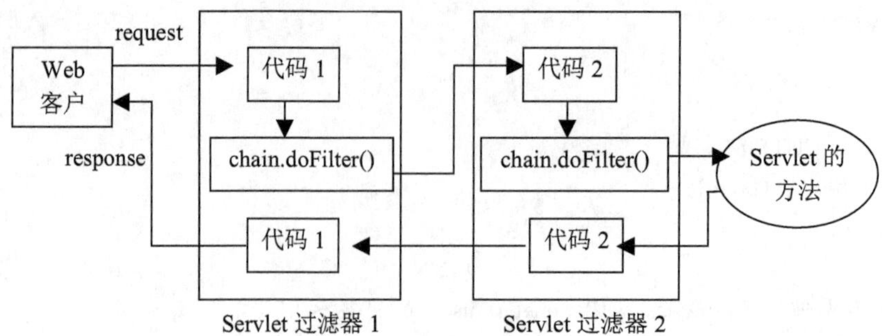

图 9-5

多个 Servlet 过滤器顺序是靠它们在 web.xml 的顺序来决定的，来看前面的一个例子，使用两个 Servlet 过滤器，第一个过滤器过滤用户没有输入用户名和发言的内容，第二个过滤器过滤用户名在黑名单上的客户。两个 Servlet 过滤器代码如示例 9-5 和示例 9-6 所示。

示例 9-5：

```
package control;

import java.io.IOException;
import java.io.PrintWriter;

import javax.servlet.Filter;
import javax.servlet.FilterChain;
import javax.servlet.FilterConfig;
import javax.servlet.ServletException;
import javax.servlet.ServletRequest;
import javax.servlet.ServletResponse;
import javax.servlet.annotation.WebFilter;
import javax.servlet.http.HttpServlet;

@WebFilter("/CheckInputFilter")
public class CheckInputFilter extends HttpServlet implements Filter {

  public CheckInputFilter() {
  }

  public void destroy() {
  }

  public void doFilter(ServletRequest request, ServletResponse response, FilterChain chain)
          throws IOException, ServletException {
     String username = request.getParameter("username");
     String content = request.getParameter("content");
     request.setCharacterEncoding("UTF-8");
     response.setContentType("text/html;charset=UTF-8");
     PrintWriter out = response.getWriter();
```

```
        if (username.length() == 0 || content.length() == 0) {
            out.print("<html><head><title>错误</title>");
            out.println("<body><h1>用户名或内容为空</h1>");
            out.println("<p><a href=input.jsp>回到主页面</a></p>");
            out.println("</body></html>");
        } else {
            chain.doFilter(request, response);
        }

    }

    public void init(FilterConfig fConfig) throws ServletException {
    }

}
```

示例 9-6：

```
package control;

import java.io.IOException;
import java.io.PrintWriter;

import javax.servlet.Filter;
import javax.servlet.FilterChain;
import javax.servlet.FilterConfig;
import javax.servlet.ServletException;
import javax.servlet.ServletRequest;
import javax.servlet.ServletResponse;
import javax.servlet.annotation.WebFilter;
import javax.servlet.http.HttpServlet;

@WebFilter("/BlacklistFilter")
public class BlacklistFilter extends HttpServlet implements Filter {
    private FilterConfig config = null;
    private String blackList = "";

    public void destroy() {
        this.config = null;
    }

    public void doFilter(ServletRequest request, ServletResponse response, FilterChain chain)
            throws IOException, ServletException {
        String username = request.getParameter("username");
        request.setCharacterEncoding("utf-8");
        response.setContentType("text/html;charset=utf-8");
        PrintWriter out = response.getWriter();
```

```
        if (username != null && username.indexOf(blackList) != -1) {
            out.println("<html><head><title>Filter 测试</title></head><body>");
            out.println("<h1>对不起, " + username + ",你没有权利使用 Web 资源</h1>");
            out.print("</body></html>");
            return;
        } else {
            chain.doFilter(request, response);
        }
    }

    @Override
    public void init(FilterConfig config) throws ServletException {
        this.config = config;
        blackList = config.getInitParameter("blacklist");
    }

}
```

web.xml 里的 Servlet 过滤器的配置如示例 9-7 所示，注意 CheckInputFilter 顺序在 BlacklistFilter 前面。

示例 9-7：

```
<?xml version="1.0" encoding="UTF-8"?>
<web-app version="2.4"
    xmlns="http://java.sun.com/xml/ns/j2ee"
    xmlns:xsi="http://www.w3.org/2001/XMLSchema-instance"
    xsi:schemaLocation="http://java.sun.com/xml/ns/j2ee
    http://java.sun.com/xml/ns/j2ee/web-app_2_4.xsd">
        <filter>
        <filter-name>CheckInputFilter</filter-name>
        <filter-class>control.CheckInputFilter</filter-class>
        </filter>
    <filter>
        <filter-name>BlacklistFilter</filter-name>
        <filter-class>control.BlacklistFilter</filter-class>
        <init-param>
            <param-name>blacklist</param-name>
            <param-value>tom</param-value>
        </init-param>
    </filter>
    <filter-mapping>
        <filter-name>CheckInputFilter</filter-name>
        <url-pattern>/show.jsp</url-pattern>
    </filter-mapping>
```

```
    <filter-mapping>
        <filter-name>BlacklistFilter</filter-name>
        <url-pattern>/show.jsp</url-pattern>
    </filter-mapping>
</web-app>
```

下面建立一个名为 input.jsp 的 JSP 文件来测试这两个 Servlet 过滤器，该 JSP 页面提交请求给 show.jsp。这两个 JSP 文件如示例 9-8 所示。

示例 9-8：

```
<%@ page language="java" contentType="text/html; charset=UTF-8"
    pageEncoding="UTF-8"%>
<!DOCTYPE html>
<html>
<head>
<meta charset="UTF-8">
<title>显示用户信息</title>
</head>
<body>
  <h1><%=new String((request.getParameter("username")).getBytes("ISO-8859-1"),"UTF-8") %></h1>
     <p>你留言的内容是:<%=new
String((request.getParameter("content")).getBytes("ISO-8859-1"),"UTF-8")
%></p>
</body>
</html>
```

完成后，把工程发布到 Tomcat 服务器中，本例的工程名为 MyNinthWebApp。发布完成后，启动 Tomcat 服务器，并在 IE 的地址栏中输入“http://localhost:8080/MyNinthWebApp/input.jsp”，显示结果如图 9-6 所示。

当用户名或者密码输入框为空时，单击“提交”按钮，显示结果如图 9-7 所示；当输入的用户名为 tom 时，显示结果如图 9-8 所示。

当用户名是 tom 以外的任何名字时，例如“李易峰”，输入内容是“大家好，我是李易峰，欢迎加入我的战队”，单击“提交”按钮，显示结果如图 9-9 所示。

图 9-6

图 9-7

图 9-8

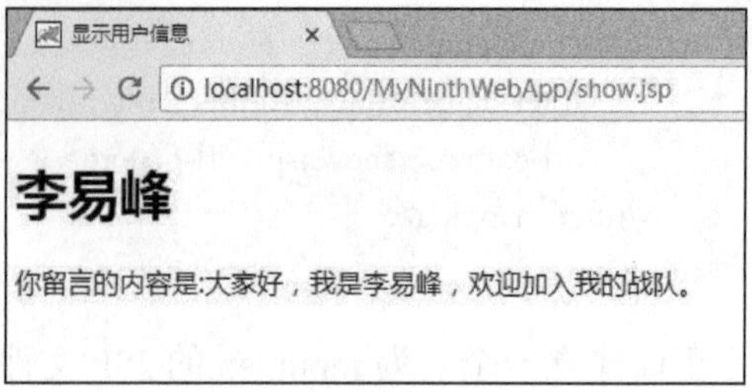

图 9-9

【单元小结】

- 过滤器是访问用户请求和 Web 资源响应的 Web 服务组件，可用来验证用户请求、加密和解密数据。
- 过滤器定义将过滤器名称和特定的类相关联。
- 过滤器映射用于将过滤器映射到 Web 资源。
- FilterChain 接口用于调用过滤器中的下一个过滤器。
- 在初始化的过程中，Servlet 容器使用 FilterConfig 接口将信息传递给过滤器。

【单元自测】

1. (　　)用于调用过滤器链中的下一个过滤器。

   A. FilterConfig()　　B. Filter()

   C. FilterChain()　　D. RequestDispatcher()

2. 每当传递请求或响应时调用 Filter 的(　　)方法。

   A. init()　　B. destroy()

   C. doFilter()　　D. getInitParameter()

3. 在初始化过程中，Servlet 容器使用(　　)将信息传递给过滤器。

   A. FilterChain　　B. FilterConfig

   C. Filter　　D. FilterName

4. 在使用过滤器之前调用(　　)方法。

   A. doFilter()　　B. destroy()

   C. init()　　D. getFilterName()

5. FilterConfig 的(　　)方法返回 init 参数的名称。

   A. getServletContext()　　B. getInitParameter(String name)

   C. getFilterName()　　D. getInitParameterNames()

【上机实战】

上机目标

掌握如何在 Web 应用中使用过滤器。

上机练习

◆ 第一阶段 ◆

练习 1：过滤 IP 地址

【问题描述】

实现一个过滤器，能实现客户端的 IP 地址的过滤，来自某个 IP 的主机不能够访问 Web 资源。

【问题分析】

本练习主要是为了巩固理论课所讲的过滤器的用法。创建一个过滤器，通过 HttpServletRequest 的 getRemoteAddress()方法取得请求中的 IP，把该 IP 和要阻止的 IP 进行比对。当来自客户端的请求中包含 IP 地址和要阻止的 IP 匹配时，禁止用户访问资源。

【参考步骤】

(1) 继续在本章节工程项目下创建一个 Servlet 过滤器，取名为 IPBlockFilter。使用下面的方法获得客户端的 IP 地址。

```
String remoteIP =request.getRemoteAddress()
```

完整代码如示例 9-9 所示。

示例 9-9：

```
package control;

import java.io.IOException;
import java.util.StringTokenizer;

import javax.servlet.Filter;
import javax.servlet.FilterChain;
import javax.servlet.FilterConfig;
import javax.servlet.ServletException;
import javax.servlet.ServletRequest;
```

```
import javax.servlet.ServletResponse;
import javax.servlet.annotation.WebFilter;
import javax.servlet.http.HttpServlet;
import javax.servlet.http.HttpServletResponse;

@WebFilter("/IPBlockFilter")
public class IPBlockFilter extends HttpServlet implements Filter {

    public FilterConfig config = null;
    public final static String IP_RANGE = "127.0";

    public IPBlockFilter() {
    }

    public void destroy() {
    }

    public void doFilter(ServletRequest request, ServletResponse response, FilterChain chain)
            throws IOException, ServletException {
        String ip = request.getRemoteAddr();
        HttpServletResponse httpResp = null;

        if (response instanceof HttpServletResponse)
            httpResp = (HttpServletResponse) response;
        StringTokenizer toke = new StringTokenizer(ip, ".");

        int dots = 0;
        String byte1 = "";
        String byte2 = "";
        String clientIP = "";

        while (toke.hasMoreTokens()) {
            ++dots;
            if (dots == 1) {
                byte1 = toke.nextToken();
            } else {
                byte2 = toke.nextToken();
                break;
            }
        }

        clientIP = byte1 + "." + byte2;

        if (IP_RANGE.equals(clientIP)) {
            httpResp.sendError(HttpServletResponse.SC_FORBIDDEN, "对不起，我们无法让你访问网
            络资源");
        } else {
```

```
            chain.doFilter(request, response);
        }

    }

    public void init(FilterConfig fConfig) throws ServletException {
        this.config = config;
    }

}
```

(2) 建立一个名为 test.jsp 的 JSP 文件，代码如示例 9-10 所示。

示例 9-10：

```
<%@ page language="java" contentType="text/html; charset=UTF-8"
    pageEncoding="UTF-8"%>
<!DOCTYPE html>
<html>
<head>
<meta charset="UTF-8">
<title>Insert title here</title>
</head>
<body>
 请求 IP 地址
</body>
</html>
```

(3) 发布该 Web 应用到 Tomcat 服务器并启动，打开 IE 浏览器，在地址栏中输入“http://127.0.0.1:8080/MyNinthWebApp/test.jsp”，显示结果如图 9-10 所示。

图 9-10

练习 2：使用过滤器限制客户访问网站的时间

【问题描述】

创建一个过滤器，限制用户只有在规定的时间段内才可以访问游戏网站的资源。

【问题分析】

创建一个过滤器，有两个<init-param>参数，分别为可以访问网站的开始和结束时间。过滤器通过 Calendar 读出用户的访问时间(小时数)，判断若为合法的时间段，则让用户访问资源，否则显示无法访问的信息。

【参考步骤】

(1) 在 Eclipse 中建立 Web 工程，取名 FilterDemo，然后建立 Servlet 过滤器，注意要继承 HttpServlet 和实现 Filter 接口，代码如示例 9-11 所示。

示例 9-11：

```
package control;

import java.io.IOException;
import java.io.PrintWriter;
import java.util.Calendar;

import javax.servlet.Filter;
import javax.servlet.FilterChain;
import javax.servlet.FilterConfig;
import javax.servlet.ServletException;
import javax.servlet.ServletRequest;
import javax.servlet.ServletResponse;
import javax.servlet.annotation.WebFilter;
import javax.servlet.http.HttpServlet;

@WebFilter("/StopUseFilter")
public class StopUseFilter extends HttpServlet implements Filter {
  private FilterConfig filterConfig;

  private int starthour = 0;

  private int stophour = 24;

  public void doFilter(ServletRequest request, ServletResponse response, FilterChain filterChain)
        throws IOException, ServletException {
     Calendar myCal = Calendar.getInstance();
     int curhour = myCal.get(Calendar.HOUR_OF_DAY);
     filterConfig.getServletContext()
            .log("in StopGamesFilter cur:" + curhour + ", start: " + starthour + ", end: " + stophour);
     try {
        if ((curhour >= stophour) || (curhour <= starthour)) {
            response.setContentType("text/html;charset=UTF-8");
            PrintWriter out = response.getWriter();
            out.println("<html><head></head><body>");
            out.println("<h1>对不起，此时不允许玩游戏!</h1>");
```

```
                out.println("</body></html>");

                out.flush();
                filterConfig.getServletContext().log("Access to game page denied");
                return;
            }
            filterConfig.getServletContext().log("Access to game page granted");
            filterChain.doFilter(request, response);
            filterConfig.getServletContext().log("Getting out of StopGamesFilter");
        } catch (ServletException sx) {
            filterConfig.getServletContext().log(sx.getMessage());
        } catch (IOException iox) {
            filterConfig.getServletContext().log(iox.getMessage());
        }
    }

    public void init(FilterConfig filterConfig) throws ServletException {
        this.filterConfig = filterConfig;
        String tpString;
        if ((tpString = filterConfig.getInitParameter("starthour")) != null) {
            starthour = Integer.parseInt(tpString, 10);
        }
        if ((tpString = filterConfig.getInitParameter("stophour")) != null) {
            stophour = Integer.parseInt(tpString, 10);
        }
    }

    public void destroy() {
    }
}
```

(2) 配置 web.xml，如示例 9-12 所示。

示例 9-12：

```
<?xml version="1.0" encoding="UTF-8"?>
<web-app version="2.4" xmlns="http://java.sun.com/xml/ns/j2ee"
    xmlns:xsi="http://www.w3.org/2001/XMLSchema-instance"
    xsi:schemaLocation="http://java.sun.com/xml/ns/j2ee
    http://java.sun.com/xml/ns/j2ee/web-app_2_4.xsd">
    <filter>
        <filter-name>StopUseFilter</filter-name>
        <filter-class>control.StopUseFilter</filter-class>
        <init-param>
            <param-name>strathour</param-name>
            <param-value>8</param-value>
        </init-param>
        <init-param>
```

```
            <param-name>stophour</param-name>
            <param-value>9</param-value>
        </init-param>
    </filter>

    <filter-mapping>
        <filter-name>StopUseFilter</filter-name>
        <url-pattern>/*</url-pattern>
    </filter-mapping>
</web-app>
```

(3) 写一个请求的测试页面，代码如示例 9-13 所示。

示例 9-13：

```
<html>
  <head>
    <title>game.html</title>
  </head>
  <body>
    欢迎参与网络游戏<br>
  </body>
</html>
```

(4) 发布 Web 应用到 Tomcat 服务器并启动，在 IE 地址栏中输入“http://127.0.0.1:8080/FilterDemo/game.html”，显示结果如图 9-11 所示。

图 9-11

◆ 第二阶段 ◆

练习 3：在 Web 应用中使用过滤器链

【问题描述】

创建一个过滤器链，检索远程主机、使用的协议、内容类型、内容长度、远程地址和用户名，在 Servlet 中显示这些信息。

【问题分析】

使用 FilterChain 接口调用过滤器链中的下一个过滤器。创建过滤器，检索详细信息并在调用最后一个过滤器时将此信息传递给 Servlet。

【参考步骤】

(1) 在 Eclipse 中建立 Web 工程，取名为 FilterTest。然后建立一个名为 RequestServlet 的 Servlet，代码如示例 9-14 所示。

示例 9-14：

```
package control;

import java.io.IOException;
import java.io.PrintWriter;

import javax.servlet.ServletException;
import javax.servlet.annotation.WebServlet;
import javax.servlet.http.HttpServlet;
import javax.servlet.http.HttpServletRequest;
import javax.servlet.http.HttpServletResponse;
import javax.servlet.http.HttpSession;

@WebServlet("/RequestServlet")
public class RequestServlet extends HttpServlet {
   private static final long serialVersionUID = 1L;

   public RequestServlet() {
      super();
   }

   protected void doGet(HttpServletRequest request, HttpServletResponse response)
            throws ServletException, IOException {
      request.setCharacterEncoding("UTF-8");
      response.setContentType("text/html;charset=UTF-8");
      PrintWriter out = response.getWriter();
      HttpSession ses = request.getSession(true);
      ses.setAttribute("ContentType", String.valueOf(request.getContentType()));
      String protocol = (String) ses.getAttribute("Protocol");
      String remotehost = (String) ses.getAttribute("RemoteHost");
      String contentlength = (String) ses.getAttribute("ContentLength");
      String servername = (String) ses.getAttribute("ServerName");
      String remoteaddr = (String) ses.getAttribute("RemoteAddr");
      String username = (String) ses.getAttribute("username");
      out.println("<html>");
      out.println("<head><title>RequestedServlet</title></head>");
      out.println("<body bgcolor=\"#ffffff\">");
      out.println("<h3>检索到的详细信息</h3>");
      out.println("<b>使用的协议：  </b>" + protocol + "<br/>");
      out.println("<b>远程主机：  </b>" + remotehost + "<br/>");
      out.println("<b>内容长度：  </b>" + contentlength + "<br/>");
      out.println("<b>服务器名称：  </b>" + servername + "<br/>");
```

```
        out.println("<b>远程地址： </b>" + remoteaddr + "<br/>");
        out.println("<b>用户名： </b>" + username + "<br/>");
        out.println("</body>");
        out.println("</html>");
        out.close();

    }

    protected void doPost(HttpServletRequest request, HttpServletResponse response)
            throws ServletException, IOException {
        doGet(request, response);
    }

}
```

(2) 在当前的工程中创建一个名字叫 FirstFilter 的 Servlet，注意继承 HttpServlet 和实现 Filter 接口，代码如示例 9-15 所示。

示例 9-15：

```
package control;

import java.io.IOException;

import javax.servlet.Filter;
import javax.servlet.FilterChain;
import javax.servlet.FilterConfig;
import javax.servlet.ServletException;
import javax.servlet.ServletRequest;
import javax.servlet.ServletResponse;
import javax.servlet.annotation.WebFilter;
import javax.servlet.http.HttpServlet;
import javax.servlet.http.HttpServletRequest;
import javax.servlet.http.HttpSession;

@WebFilter("/FirstFilter")
public class FirstFilter extends HttpServlet implements Filter {
    private FilterConfig filterConfig;

    public FirstFilter() {
    }

    public void destroy() {
    }

    public void doFilter(ServletRequest request, ServletResponse response, FilterChain chain)
            throws IOException, ServletException {
```

```
        HttpSession ses = ((HttpServletRequest) request).getSession(true);
        ses.setAttribute("Protocol", String.valueOf(request.getProtocol()));
        ses.setAttribute("RemoteHost", request.getRemoteHost());
        try {
            chain.doFilter(request, response);
        } catch (ServletException sx) {
            filterConfig.getServletContext().log(sx.getMessage());
        } catch (IOException iox) {
            filterConfig.getServletContext().log(iox.getMessage());
        }

    }

    public void init(FilterConfig fConfig) throws ServletException {
        this.filterConfig = filterConfig;
    }

}
```

(3) 在当前的工程中创建第二个名字为 SecondFilter 的 Servlet，注意继承 HttpServlet 和实现 Filter 接口，代码如示例 9-16 所示。

示例 9-16：

```
package control;

import java.io.IOException;

import javax.servlet.Filter;
import javax.servlet.FilterChain;
import javax.servlet.FilterConfig;
import javax.servlet.ServletException;
import javax.servlet.ServletRequest;
import javax.servlet.ServletResponse;
import javax.servlet.annotation.WebFilter;
import javax.servlet.http.HttpServlet;
import javax.servlet.http.HttpServletRequest;
import javax.servlet.http.HttpSession;

@WebFilter("/SecondFilter")
public class SecondFilter extends HttpServlet implements Filter {
    private FilterConfig filterConfig;

    public SecondFilter() {
    }

    public void destroy() {
    }
```

```
    public void doFilter(ServletRequest request, ServletResponse response, FilterChain chain)
            throws IOException, ServletException {
        HttpSession ses = ((HttpServletRequest) request).getSession(true);
        ses.setAttribute("ContentLength", String.valueOf(request.getContentLength()));
        try {
            chain.doFilter(request, response);
        } catch (ServletException sx) {
            filterConfig.getServletContext().log(sx.getMessage());
        } catch (IOException iox) {
            filterConfig.getServletContext().log(iox.getMessage());
        }

    }

    public void init(FilterConfig fConfig) throws ServletException {
        this.filterConfig = filterConfig;
    }

}
```

(4) 在当前的工程中创建第三个名字为 ThirdFilter 的 Servlet，注意继承 HttpServlet 和实现 Filter 接口，代码如示例 9-17 所示。

示例 9-17：

```
package control;

import java.io.IOException;

import javax.servlet.Filter;
import javax.servlet.FilterChain;
import javax.servlet.FilterConfig;
import javax.servlet.ServletException;
import javax.servlet.ServletRequest;
import javax.servlet.ServletResponse;
import javax.servlet.annotation.WebFilter;
import javax.servlet.http.HttpServlet;
import javax.servlet.http.HttpServletRequest;
import javax.servlet.http.HttpSession;

@WebFilter("/ThirdFilter")
public class ThirdFilter extends HttpServlet implements Filter {
    private FilterConfig filterConfig = null;

    public ThirdFilter() {
    }
```

```
    public void destroy() {
    }

    public void doFilter(ServletRequest request, ServletResponse response, FilterChain chain)
            throws IOException, ServletException {
        HttpSession ses = ((HttpServletRequest) request).getSession(true);
        ses.setAttribute("ServerName", request.getServerName());
        ses.setAttribute("RemoteAddr", request.getRemoteAddr());
        String username = request.getParameter("username");
        ses.setAttribute("username", username);
        try {
            chain.doFilter(request, response);
        } catch (ServletException sx) {
            filterConfig.getServletContext().log(sx.getMessage());
        } catch (IOException iox) {
            filterConfig.getServletContext().log(iox.getMessage());
        }

    }

    public void init(FilterConfig fConfig) throws ServletException {
        this.filterConfig = filterConfig;
    }

}
```

(5) 在当前的工程中创建一个名称为 requestuser.html 的 HTML 网页，接受用户名并调用当前工程中的 Servlet，代码如示例 9-18 所示。

示例 9-18：

```
<!DOCTYPE html>
<html>
<head>
<meta charset="UTF-8">
<title>简单的过滤器示例</title>
</head>
<body>
<h1>过滤器链的示例。</h1>
<form action="servlet/RequestServlet">
<b>请输入您的姓名：</b>
<br/>
<input type="text" name="username"/>
<input type="submit"/>
</form>
</body>
</html>
```

(6) 配置 web.xml，代码如示例 9-19 所示。

示例 9-19：

```
<?xml version="1.0" encoding="UTF-8"?>
<web-app xmlns:xsi="http://www.w3.org/2001/XMLSchema-instance"
xmlns="http://java.sun.com/xml/ns/j2ee" xmlns:web="http://xmlns.jcp.org/xml/ns/javaee"
xsi:schemaLocation="http://java.sun.com/xml/ns/j2ee http://java.sun.com/xml/ns/j2ee/web-app_2_4.xsd
http://xmlns.jcp.org/xml/ns/javaee http://java.sun.com/xml/ns/javaee/web-app_2_5.xsd" version="2.4">
    <servlet>
        <servlet-name>RequestServlet</servlet-name>
        <servlet-class>control.RequestServlet</servlet-class>
    </servlet>
    <filter>
        <filter-name>FirstFilter</filter-name>
        <filter-class>control.FirstFilter</filter-class>
    </filter>
    <filter>
        <filter-name>SecondFilter</filter-name>
        <filter-class>control.SecondFilter</filter-class>
    </filter>
    <filter>
        <filter-name>ThirdFilter</filter-name>
        <filter-class>control.ThirdFilter</filter-class>
    </filter>

    <servlet-mapping>
        <servlet-name>RequestServlet</servlet-name>
        <url-pattern>/servlet/RequestServlet</url-pattern>
    </servlet-mapping>
    <filter-mapping>
        <filter-name>FirstFilter</filter-name>
        <url-pattern>/*</url-pattern>
    </filter-mapping>
    <filter-mapping>
        <filter-name>SecondFilter</filter-name>
        <url-pattern>/*</url-pattern>
    </filter-mapping>
    <filter-mapping>
        <filter-name>ThirdFilter</filter-name>
        <url-pattern>/*</url-pattern>
    </filter-mapping>
</web-app>
```

(7) 发布 Web 应用到 Tomcat 服务器并启动，在 IE 地址栏中输入“http://localhost:8080/FilterTest/requestuser.html”，显示结果如图 9-12 所示。

图 9-12

(8) 在“请输入您的姓名”文本框中输入一个姓名，如 tom，然后单击“提交”按钮，显示结果如图 9-13 所示。

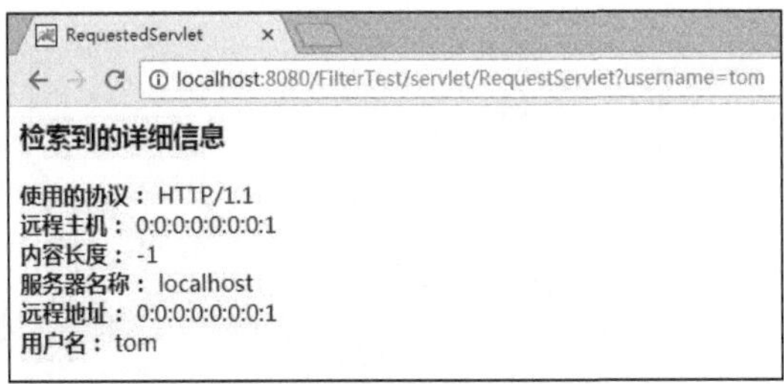

图 9-13

【拓展作业】

1. 编写一个 Servlet 过滤器解决 JSP 页面的中文问题。

2. 编写一个过滤器用于过滤一个 Web 项目。凡是不在根目录中的资源都必须在用户登录后才能访问，未登录的用户访问非根目录资源将直接转向到登录页面。用户的信息放在数据库的表中。

单元十 监听器

课程目标

- ► 理解 Servlet 监听器
- ► 理解 Servlet 监听器的执行流程
- ► 掌握如何实现简单的 Servlet 监听器

简 介

Servlet 监听器就是一个实现了特定接口的 Java 类，这个 Java 类用于监听另一个 Java 类的方法调用或属性的改变。当被监听对象发生上述事件后，监听器中指定的方法将会被立即执行。

10.1 监听器入门

在监听器中有这些术语：事件源、监听器、事件源和监听器绑定、事件。下面以汽车报警器为例来说明：事件源指的是被监听的对象，也就是汽车；监听器指的是监听的对象，在这里也就是报警器；事件源和监听器绑定就是在汽车上安装报警器；事件指的是事件源对象的改变，例如有人踹了汽车这个事件从而引发报警器报警，它的主要功能是获得事件源对象。

下面不使用 Servlet 监听器，而是自己编写一个监听器小示例作为入门演示。

(1) 新建一个 Person 实体类和一个 PersonEvent 类，代码如示例 10-1 和示例 10-2 所示。

示例 10-1：

```
package domain;

import listener.IPersonRunListener;

public class Person {
    private String name;
    private IPersonRunListener listener;

    public Person() {
        super();
        // TODO Auto-generated constructor stub
    }

    public String getName() {
        return name;
    }

    public void setName(String name) {
        this.name = name;
    }

    public IPersonRunListener getListener() {
        return listener;
```

```
    }

    public void setListener(IPersonRunListener listener) {
        this.listener = listener;
    }

    public void run() {
        System.out.println(name + "：开始跑了..");
        if (listener != null) {
                listener.fighting(new PersonEvent(this));
        }
    }

    public void addPersonListener(IPersonRunListener listener) {
        this.listener = listener;
    }

    public Person(String name) {
        super();
        this.name = name;

    }

    @Override
    public String toString() {
        return "Person [name=" + name + ", listener=" + listener + "]";
    }
}
```

示例 10-2：

```
package domain;

public class PersonEvent {
    Person p = null;

    public PersonEvent(Person p) {
        this.p = p;
    }

    public String getName() {
        return p.getName();
    }

    public Object getSource() {
        return p;
    }
```

```
}
```

(2) 编写一个 IPersonRunListener 接口，以及一个默认的实现类 DefaultCatListener，代码如示例 10-3 和示例 10-4 所示。

示例 10-3：

```
package listener;

import domain.PersonEvent;

public interface IPersonRunListener {
    public void fighting(PersonEvent pe);
}
```

示例 10-4：

```
package listener;

import domain.PersonEvent;

public class DefaultCatListener implements IPersonRunListener {

    @Override
    public void fighting(PersonEvent pe) {
        System.out.println("默认监听器的实现类上线了，开始监测他的跑步行为。");
    }

}
```

(3) 编写 Demo 类，如示例 10-5 所示。

示例 10-5：

```
package control;

import domain.Person;
import domain.PersonEvent;
import listener.DefaultCatListener;
import listener.IPersonRunListener;

public class Demo {
    public static void main(String[] args) {
        Person p1 = new Person("张三");
        Person p2 = new Person("Jack");
//使用匿名内部类实现监听器接口方法
        IPersonRunListener listener = new IPersonRunListener() {
```

```
            @Override
            public void fighting(PersonEvent pe) {
                System.out.println(pe.getSource() + "监测到他已经跑完了...");
                if (pe.getName().equals("张三")) {
                    System.out.println(pe.getName() + "跑到了第一名...");
                }
            }
        };
        p1.addPersonListener(listener); // 在事件源上绑定监听器
        p2.addPersonListener(listener);
        p1.run();
        p2.run();

        Person p3 = new Person("李四");
        p3.addPersonListener(new DefaultCatListener());// 在事件源上绑定默认监听器
        p3.run();
    }
}
```

(4) 运行编写的 Demo 类，查看控制台结果，如图 10-1 所示。

```
Markers  Properties  Servers  Data Source Explorer  Snippets  Console  Maven Repositories  Terminal
<terminated> Demo [Java Application] D:\Program Files\Java\jdk1.8.0_161\bin\javaw.exe (2018年9月14日 上午10:56:21)
张三：开始跑了..
Person [name=张三, listener=control.Demo$1@70dea4e]监测到他已经跑完了...
张三跑到了第一名...
Jack：开始跑了..
Person [name=Jack, listener=control.Demo$1@70dea4e]监测到他已经跑完了...
李四：开始跑了..
默认监听器的实现类上线了，开始监测他的跑步行为。
```

图 10-1

10.2 监听器执行流程

监听器的执行类似于触发器，当某些动作(接口中定义好的)执行时就会触发相应的 Listener，执行相应的操作。如果没有执行相应的动作，则监听器就一直监听着，没有操作。下面根据示例 10-1，用时序图 10-2 来展示它的执行流程。

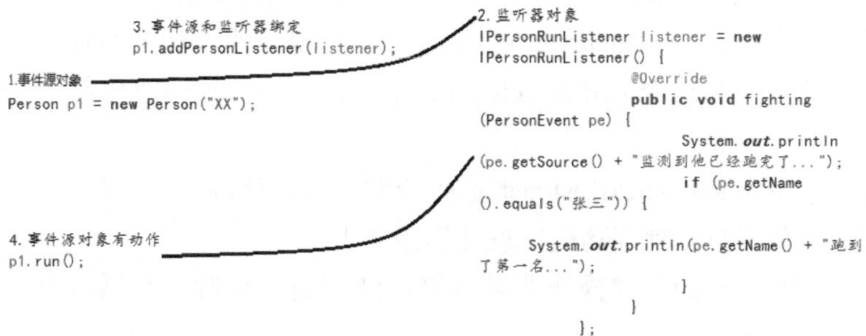

图 10-2

从图中描述可以更加清晰地了解到事件源和监听器绑定后，监听器就在时刻监听着，如果事件源中对应的动作有执行，就会触发相对应的方法。下面我们将详细讲解 Servlet 中的监听器，了解常用的三类监听器。

10.3 Servlet 中的监听器

在 Servlet API 中提供了大量监听器接口来帮助开发者实现对 Web 应用内特定事件进行监听，从而当 Web 应用内这些特定事件发生时，回调监听器内的事件监听方法来实现我们所需要的特殊功能。

Web 容器使用不同的监听器接口来实现对不同事件的监听，常用的 Web 事件监听器接口可分为以下三类。

- 与 Servlet 上下文相关的监听器接口，如 ServletContextListener 接口。
- 与会话相关的监听器接口，如 HttpSessionListener 接口。
- 与请求相关的监听器接口，如 ServletRequestListener 接口。

下面对这三个监听器接口进行一一讲解。

10.3.1 ServletContextListener 监听器

ServletContextListener 接口用于监听 Web 应用程序的 ServletContext 对象的创建和销毁事件。每个 Web 应用对应一个 ServletContext 对象，在 Web 容器启动时创建，在容器关闭时销毁。当 Web 应用程序中声明了一个实现 ServletContextListener 接口的事件监听器后，Web 容器在创建或销毁时就会产生一个 ServletContextEvent 事件对象，然后再执行监听器中的相应事件处理方法，并将 ServletContextEvent 事件对象传递给这些方法。在 ServletContextListener 接口中定义了如下两个事件处理方法。

- contextInitialized(ServletContextEvent sce)：当 ServletContext 对象被创建时，Web 容器将调用此方法。该方法接收 ServletContextEvent 事件对象，通过此对象可获得当前被创建的 ServletContext 对象。
- contextDestroyed(ServletContextEvent sce)：当 ServletContext 对象被销毁时，Web 容器调用此方法，同时向其传递 ServletContextEvent 事件对象。

上述处理方法中，ServletContextEvent 为一个事件类，用于通知 Web 应用程序中上下文对象的改变，该类具有一个 getServletContext()方法，用于返回改变前的 ServletContext 对象。

下面编写一个 ServletContextListener 监听器的案例，该案例实现对一个保存在应用域属性中的访问计数值的持久保存功能。设计思想如下：

(1) 由于计数器数值的存取操作非常频繁，通常将其保存在容器内存中的应用域属性中。

(2) 在 Web 应用终止时，把保存在应用域属性中的计数器数值永久性地保存到一个文件中。

(3) 在 Web 应用启动时从文件中读取计数器的数值，并将其存入应用域属性中。代码如示例 10-6 所示。

示例 10-6：

```
package listener;

import java.io.BufferedReader;
import java.io.IOException;
import java.io.InputStreamReader;
import java.io.PrintWriter;

import javax.servlet.ServletContext;
import javax.servlet.ServletContextEvent;
import javax.servlet.ServletContextListener;
import javax.servlet.annotation.WebListener;

@WebListener
public class VisitCountListener implements ServletContextListener {

public VisitCountListener() {
}

/**
 * Web 应用停止时，容器调用此方法
 */

public void contextDestroyed(ServletContextEvent sce) {
        // 获取 ServletContext 对象
        ServletContext context = sce.getServletContext();
        // 输出应用范围获得计数器对象
        Integer counter = (Integer) context.getAttribute("count");
        if (counter != null) {
                try {
                        // 把计数器的数值写到项目发布目录下的 count.txt 文件中
                        context.log(context.getRealPath("/"));
                        String filepath = context.getRealPath("/") + "/count.txt";

                        PrintWriter pw = new PrintWriter(filepath);
                        pw.println(counter.intValue());
                        pw.close();

                } catch (IOException e) {
                        e.printStackTrace();
                }
        }
```

```
    }

    /**
     * Web 应用初始化时，容器调用此方法
     */

    public void contextInitialized(ServletContextEvent sce) {
        // 获取 ServletContext 对象
        ServletContext context = sce.getServletContext();
        // 输出应用初始化日志信息
        context.log(context.getServletContextName() + "应用开始初始化");
        try {
            // 从文件中读取计数器的数值
            BufferedReader reader = new BufferedReader(
                    new InputStreamReader(context.getResourceAsStream("/count.txt")));
            String strcount = reader.readLine();
            if (strcount == null || "".equals(strcount)) {
                strcount = "0";
            }
            int count = Integer.parseInt(strcount);
            reader.close();
            // 把计数器对象保存到 Web 应用范围
            context.setAttribute("count", count);

        } catch (IOException e) {
            e.printStackTrace();
        }

    }

}
```

下面再编写一个 Servlet 来测试效果，代码如示例 10-7 所示。

示例 10-7：

```
package control;

import java.io.IOException;
import java.io.PrintWriter;

import javax.servlet.ServletContext;
import javax.servlet.ServletException;
import javax.servlet.annotation.WebServlet;
import javax.servlet.http.HttpServlet;
import javax.servlet.http.HttpServletRequest;
import javax.servlet.http.HttpServletResponse;
```

```
@WebServlet("/ContextAttributeServlet")
public class ContextAttributeServlet extends HttpServlet {
  private static final long serialVersionUID = 1L;

  public ContextAttributeServlet() {
      super();
  }

  protected void doGet(HttpServletRequest request, HttpServletResponse response)
          throws ServletException, IOException {
      // 设置响应到客户端的文本类型
      response.setContentType("text/html;charset=utf-8");
      // 获取 ServletContext 对象
      ServletContext context = super.getServletContext();
      // 从 ServletContext 对象获取 count 属性存储的计数值
      Integer count = (Integer) context.getAttribute("count");
      if (count == null) {
          count = 1;
      } else {
          count++;
      }
      // 将更新后的数值存储到 ServletContext 对象的 count 属性中
      context.setAttribute("count", count);
      // 获取输出流
      PrintWriter out = response.getWriter();
      // 输出计数信息
      out.println("<p>本地请求地址目前访问人数是：" + count + "</p>");
  }

  protected void doPost(HttpServletRequest request, HttpServletResponse response)
          throws ServletException, IOException {
      doGet(request, response);
  }

}
```

启动服务器，在浏览器中访问 http://localhost:8080/MyTenthWebApp/ContextAttributeServlet，每次刷新请求，可以看到访问人数在变化。当停止服务器后，可以看到访问人数已经持久化到本地硬盘项目根目录下了。再次启动服务器则又会重新读取本地 count.txt 上的人数值，如图 10-3 和图 10-4 所示。

图 10-3

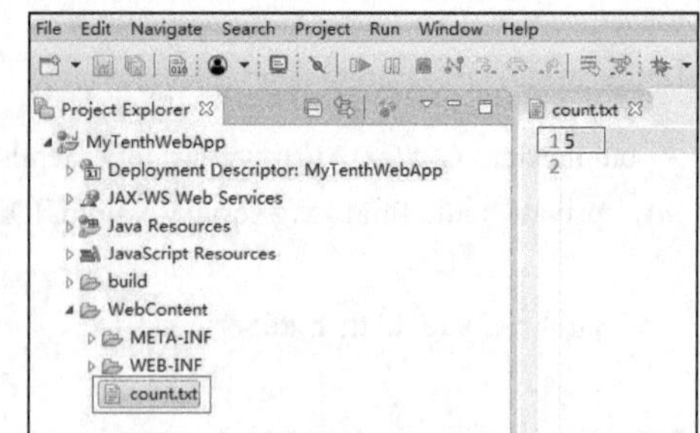

图 10-4

10.3.2 HttpSessionListener 监听器

HttpSessionListener 接口用于监听用户会话对象 HttpSession 的创建和销毁事件。每个浏览器与服务器的会话状态分别对应一个 HttpSession 对象，每个 HttpSession 对象在浏览器开始与服务器会话时创建，在浏览器与服务器结束会话时销毁。当 Web 应用程序中声明了一个实现 HttpSessionListener 接口的事件监听器后，Web 容器在创建或销毁时就会产生一个 HttpSessionEvent 事件对象，然后再执行监听器中的相应事件处理方法，并将 HttpSessionEvent 事件对象传递给这些方法。在 HttpSessionListener 接口中定义了如下两个事件处理方法。

- sessionCreated(HttpSessionEvent se)：当 HttpSession 对象被创建时，Web 容器将调用此方法。该方法接收 HttpSessionEvent 事件对象，通过此对象可获得当前被创建的 HttpSession 对象。
- sessionDestroyed(HttpSessionEvent se)：当 HttpSession 对象被销毁时，Web 容器调用此方法，同时向其传递 HttpSessionEvent 事件对象。

上述处理方法中，HttpSessionEvent 为一个事件类，用于通知 Web 应用程序中会话对象的改变，该类具有一个 getSession()方法，用于返回改变前的 HttpSession 对象。

下面代码示例 10-8 演示一个实现 HttpSessionListener 接口的监听器。该案例实现对当前在线人数的统计功能。

示例 10-8：

```
package listener;

import javax.servlet.ServletContext;
import javax.servlet.annotation.WebListener;
import javax.servlet.http.HttpSessionEvent;
import javax.servlet.http.HttpSessionListener;

/**
 * 统计在线用户数量
 *
```

```
 * @author zl
 *
 */
@WebListener
public class OnlineUserNumberListener implements HttpSessionListener {
// 统计在线人数
 private int num;

 public OnlineUserNumberListener() {
 }

/**
 * 会话创建时的监听方法
 */
 public void sessionCreated(HttpSessionEvent se) {
      // 会话创建时，人数加 1
      num++;
      ServletContext context = se.getSession().getServletContext();
      // 将在线人数存入应用域属性
      context.setAttribute("onlineUserNum", num);
 }

 /**
 * 会话销毁时的监听方法
 */
 public void sessionDestroyed(HttpSessionEvent se) {
      // 会话销毁时，人数减 1
      num--;
      ServletContext context = se.getSession().getServletContext();
      // 将在线人数存入应用域属性
      context.setAttribute("onlineUserNum", num);
 }

}
```

在线用户数量显示页面代码如示例 10-9 所示。

示例 10-9：

```
<%@ page language="java" contentType="text/html; charset=UTF-8"
     pageEncoding="UTF-8"%>
<!DOCTYPE html>
<html>
<head>
<meta charset="UTF-8">
<title>在线人数统计</title>
</head>
<body>
```

```
<p>当前在线人数为：${applicationScope.onlineUserNum }</p>
<a href="logout.jsp">安全退出</a>
</body>
</html>
```

用户安全退出页面代码如示例 10-10 所示。

示例 10-10：

```
<%@ page language="java" contentType="text/html; charset=UTF-8"
    pageEncoding="UTF-8"%>
<!DOCTYPE html>
<html>
<head>
<meta charset="UTF-8">
<title>用户退出</title>
</head>
<body>
<%session.invalidate(); %>//本次会话对象失效
<p>您已经退出本系统！</p>
</body>
</html>
```

启动服务器，同时打开多个不同浏览器窗口(这里开启了 3 个)，访问 http://localhost:8080/MyTenthWebApp/onlineUserNum.jsp，然后将其中一个窗口进行“安全退出”操作，再次刷新另一个浏览器页面，会发现数量减少 1，分别如图 10-5 和图 10-6 所示。

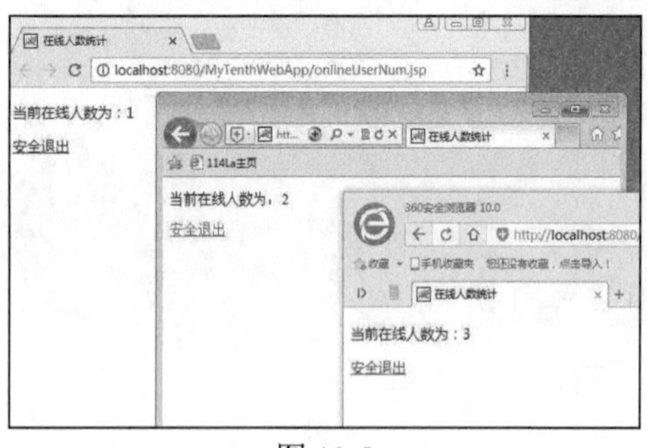

图 10-5

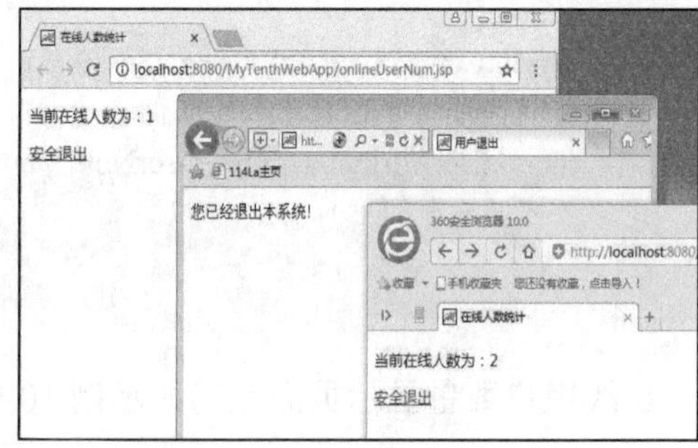

图 10-6

10.3.3 ServletRequestListener 监听器

ServletRequestListener 接口用于监听 ServletRequest 对象的创建和销毁事件。每个浏览器与服务器的会话状态分别对应一个 ServletRequest 对象，每个 ServletRequest 对象在每次访问请求时创建，在每次访问请求结束后销毁。当 Web 应用程序中声明了一个实现 ServletRequestListener 接口的事件监听器后，Web 容器在创建或销毁时就会产生一个 ServletRequestEvent 事件对象，然后将其传递给监听器中的相应事件处理方法。在

ServletRequestListener 接口中定义了如下两个事件处理方法。

- requestInitialized(ServletRequestEvent sre)：当 ServletRequest 对象被创建时，Web 容器将调用此方法。该方法接收 ServletRequestEvent 事件对象，通过此对象可获得当前被创建的 ServletRequest 对象。
- requestDestroyed(ServletRequestEvent sre)：当 ServletRequest 对象被销毁时，Web 容器调用此方法，同时向其传递 ServletRequestEvent 事件对象。

上述处理方法中，ServletRequestEvent 为一个事件类，用于通知 Web 应用程序中 ServletRequest 对象的改变，该类具有一个 getServletRequest()方法，用于返回改变前的 ServletRequest 对象。

下面代码示例 10-11 演示一个实现 ServletRequestListener 接口的监听器。该案例用来获取请求访问的资源地址、请求用户名称(若未登录名称为“游客”)、请求用户的 IP 和请求时间。

示例 10-11：

```
package listener;

import javax.servlet.ServletRequestEvent;
import javax.servlet.ServletRequestListener;
import javax.servlet.annotation.WebListener;
import javax.servlet.http.HttpServletRequest;

import bean.User;

/**
 * 获取请求访问的资源地址、请求用户名称(若未登录名称为“游客”)、请求用户的 IP 和请求时间
 *
 * @author zl
 *
 */
@WebListener
public class UserRequestInfoListener implements ServletRequestListener {

  public UserRequestInfoListener() {
  }

  /**
  * 请求结束时，容器调用此方法
  */
  public void requestDestroyed(ServletRequestEvent sre) {
  }

 /**
  * 请求初始化时，容器调用此方法
  */
```

```
public void requestInitialized(ServletRequestEvent sre) {
    // 获取 HttpServletRequest 对象
    HttpServletRequest request = (HttpServletRequest) sre.getServletRequest();
    // 获取请求用户 IP 地址
    String userIP = request.getRemoteAddr();
    // 获取请求资源地址
    String requestURI = request.getRequestURI();
    // 获取已登录请求用户名
    User user = (User) request.getSession().getAttribute("SESSION_USER");
    // 若未登录，设置请求用户名为“游客”
    String username = (user == null) ? "游客" : user.getEmail();
    StringBuffer sb = new StringBuffer();
    sb.append("本次请求访问信息：");
    sb.append("用户名称：");
    sb.append(username);
    sb.append(";用户 IP：");
    sb.append(userIP);
    sb.append(";请求地址");
    sb.append(requestURI);
    request.getServletContext().log(sb.toString());

}

}
```

启动服务器，访问 http://127.0.0.1:8080/MyTenthWebApp/，运行结果如图 10-7 所示。

```
Markers  Properties  Servers  Data Source Explorer  Snippets  Console  Maven Repositories  Terminal
Tomcat v8.5 Server at localhost [Apache Tomcat] D:\Program Files\Java\jdk1.8.0_161\bin\javaw.exe (2018年9月11日 下午5:22:35)
九月 11, 2018 5:28:08 下午 org.apache.catalina.core.ApplicationContext log
信息: 本次请求访问信息：用户名称：游客;用户IP：127.0.0.1;请求地址/MyTenthWebApp/
```

图 10-7

【单元小结】

- Servlet API 提供了大量监听器接口来帮助开发者实现对 Web 应用内特定事件进行监听。
- ServletContextListener 接口用于监听代表 Web 应用程序的 ServletContext 对象的创建和销毁事件。
- HttpSessionListener 接口用于监听用户会话对象 HttpSession 的创建和销毁事件。
- ServletRequestListener 接口用于监听 ServletRequest 对象的创建和销毁事件。

【单元自测】

1. (　　)用于监听用户会话对象。

A. HttpSessionListener　　　　B. ServletContextListener

C. ServletRequestListener　　D. ServletRequestAttributeListener

2. 在一个 Web 应用程序的整个运行周期内，Web 容器会创建和销毁 3 个重要的对象(　　)。

A. ServletContext　　B. HttpSession

C. ServletRequest　　D. ServletResponse

3. 在 ServletContextListener 接口中定义了两个事件处理方法，分别是(　　)。

A. contextInitialized()　　B. contextDestroyed()

C. requestInitialized()　　D. requestDestroyed()

4. 在 HttpSessionListneter 接口中共定义了两个事件处理方法，分别是(　　)。

A. sessionCreated()　　B. sessionDestroyed()

C. contextInitialized()　　D. contextDestroyed()

5. 监听器的部署在 web.xml 文件中配置，在配置文件中，它的位置应该在(　　)。

A. 过滤器的后面 Servlet 的前面　　B. 过滤器的后面 Servlet 的后面

C. 过滤器的前面 Servlet 的前面　　D. 过滤器的前面 Servlet 的后面

【上机实战】

上机目标

掌握如何在 Web 应用中使用监听器。

上机练习

◆ 第一阶段 ◆

练习 1：监听爱游网站访问次数

【问题描述】

实现一个监听器，能实现对爱游网站的访问次数统计并展示所有当前在线用户名称。

【问题分析】

本练习主要是为了巩固理论课所讲的监听器的用法。创建一个监听器，通过 ServletContextListener 的 contextInitialized()方法可以监听到服务器的启动，此时，读取数据库表 access_count 中的 count 字段，contextDestroyed()方法可以在服务器关闭时将最新的访问次数更新到数据库中。

【参考步骤】

(1) 创建一个 Web 工程，取名为 ChapterTenListener。

(2) 再创建一个 Servlet 监听器，取名为 ContextListener，代码如示例 10-12 所示。

示例 10-12：

```
package listener;

import java.sql.ResultSet;
import java.sql.SQLException;

import javax.servlet.ServletContext;
import javax.servlet.ServletContextEvent;
import javax.servlet.ServletContextListener;
import javax.servlet.annotation.WebListener;

import dao.CountDao;

/**
 * 监听器
 *
 */
@WebListener
public class ContextListener implements ServletContextListener {
 private CountDao dao = new CountDao();

 @Override
 public void contextDestroyed(ServletContextEvent sce) {
     // 当服务器关闭时，把 ServletContext 中网站访问次数的属性写入数据库

     ServletContext ctx = sce.getServletContext();
     int count = 0;
     if (ctx.getAttribute("access_count") != null) {
          count = Integer.parseInt(ctx.getAttribute("access_count").toString());
     }
     // update access_count set count='1';
     dao.updateCount(count);
 }

 @Override
 public void contextInitialized(ServletContextEvent sce) {
     // 当服务器启动时，从数据库中读取当前网站的访问次数
     ServletContext ctx = sce.getServletContext();
     ResultSet rs = dao.getCount();
     try {
          if (rs.next()) {
               int count = Integer.parseInt(rs.getString("count").toString());
               ctx.setAttribute("access_count", count);
          } else {
               ctx.setAttribute("access_count", 0);
```

```
                System.out.println("ServletContext init....");
            }
        } catch (SQLException e) {

            e.printStackTrace();
        } finally {
            if (rs != null) {
                try {
                    rs.close();
                } catch (SQLException e) {
                    e.printStackTrace();
                }
                rs = null;
            }

        }

    }
}
```

(3) 在 domain 包下新建 User 实体类，代码如示例 10-13 所示。

示例 10-13：

```
package domain;

import java.util.Date;

public class User {
 private Integer id;
 private String userName;
 private String password;
 private Date loginDate;

 public Date getLoginDate() {
     return loginDate;
 }

 public void setLoginDate(Date loginDate) {
     this.loginDate = loginDate;
 }

 public User() {
 }

 public Integer getId() {
     return id;
 }
```

```
    public void setId(Integer id) {
        this.id = id;
    }

    public String getUserName() {
        return userName;
    }

    public void setUserName(String userName) {
        this.userName = userName;
    }

    public String getPassword() {
        return password;
    }

    public void setPassword(String password) {
        this.password = password;
    }

    public User(String userName, String password) {
        super();
        this.userName = userName;
        this.password = password;
    }

    public User(Integer id, String userName, String password) {
        super();
        this.id = id;
        this.userName = userName;
        this.password = password;
    }

}
```

(4) 编写 dao 层 UserDao 和 CountDao，UserDao 用于查询用户登录账号和密码是否匹配以及更新本次登录时间。CountDao 用于将在线人数存储到数据库中。代码如示例 10-14 和示例 10-15 所示。

示例 10-14：

```
package dao;

import java.sql.Connection;
import java.sql.PreparedStatement;
import java.sql.ResultSet;
import java.sql.SQLException;
```

```
import domain.User;
import util.DBHelper;

public class UserDao {
//      登录
//      判断用户名和用户密码是否匹配
    public int userLogin(User user) {
        Connection conn = null;
        PreparedStatement preStmt = null;
        ResultSet userSet = null;

        try {
            conn = DBHelper.getConnection();
            String sql = "select id from user_info where userName=? and password=?";
            preStmt = conn.prepareStatement(sql);
            preStmt.setString(1, user.getUserName());
            preStmt.setString(2, user.getPassword());
            userSet = preStmt.executeQuery();
            if (userSet.next()) {
                return userSet.getInt("id");
            }
            return 0;
        } catch (Exception ex) {
            ex.printStackTrace();
            return 0;
        } finally {
            // 【 4. 必要的关闭 ResultSet、Statement 】
            if (userSet != null) {
                try {
                    userSet.close();
                    userSet = null;
                } catch (SQLException e) {
                    e.printStackTrace();
                }
            }

            if (preStmt != null) {
                try {
                    preStmt.close();
                    preStmt = null;
                } catch (SQLException e) {
                    e.printStackTrace();
                }
            }
        }
    }
```

```
    public void updateTime(User user) {
        Connection conn = null;
        PreparedStatement preStmt = null;
        ResultSet userSet = null;
        System.out.println("更新在线时间");
        try {
            conn = DBHelper.getConnection();
            String sql = "update user_info set loginDate=?";// 修改数据库中 count 字段的数值
            preStmt = conn.prepareStatement(sql);
            preStmt.setDate(1, (new java.sql.Date(user.getLoginDate().getTime())));
            int affectedRowNum = preStmt.executeUpdate();
            if (affectedRowNum > 0) {
                System.out.println("修改数据库数据成功！ ");
            }
        } catch (

        Exception ex) {
            ex.printStackTrace();
        } finally {
            if (preStmt != null) {
                try {
                    preStmt.close();
                    preStmt = null;
                } catch (SQLException e) {
                    e.printStackTrace();
                }
            }
        }
    }
}
```

示例 10-15：

```
package dao;

import java.sql.Connection;
import java.sql.PreparedStatement;
import java.sql.ResultSet;
import java.sql.SQLException;

import util.DBHelper;

public class CountDao {
    public void updateCount(int count) {
        Connection conn = null;
        PreparedStatement preStmt = null;
```

```
        ResultSet userSet = null;
        System.out.println("添加在线人数");
        try {
            conn = DBHelper.getConnection();
            String sql = "update access_count set count=?";// 修改数据库中 count 字段的数值
            preStmt = conn.prepareStatement(sql);
            preStmt.setInt(1, count);
            int affectedRowNum = preStmt.executeUpdate(sql);
            if (affectedRowNum > 0) {
                System.out.println("修改数据库数据成功！");
            }
        } catch (

        Exception ex) {
            ex.printStackTrace();
        } finally {
            if (preStmt != null) {
                try {
                    preStmt.close();
                    preStmt = null;
                } catch (SQLException e) {
                    e.printStackTrace();
                }
            }
        }
    }

    public ResultSet getCount() {

        Connection conn = null;
        PreparedStatement preStmt = null;
        ResultSet rs = null;
        try {
            conn = DBHelper.getConnection();
            String sql = "select count from access_count";
            preStmt = conn.prepareStatement(sql);
            rs = preStmt.executeQuery();
            return rs;
        } catch (Exception ex) {
            ex.printStackTrace();
            return null;

        }
    }
}
```

(5) 编写 login.jsp 登录页面，以及 LoginServlet 登录实现代码，如示例 10-16 和示例 10-17

所示。

示例 10-16：

```
<%@page import="util.CookieEncryptTool"%>
<%@page import="org.apache.jasper.tagplugins.jstl.core.ForEach"%>
<%@ page language="java" contentType="text/html; charset=UTF-8"
    pageEncoding="UTF-8"%>
    <%@ taglib prefix="c" uri="http://java.sun.com/jsp/jstl/core" %>
    <%@page import="java.util.*" %>
<!DOCTYPE html>
<html>
<head>
<meta charset="UTF-8">
<title>登录</title>
</head>
<body>
<%
String userName="";
String password="";
//从客户端读取 Cookie
Cookie[] cookies=request.getCookies();
if(cookies!=null){
 for(Cookie cookie:cookies){
    if("COOKIE_USERNAME".equals(cookie.getName())){
        //解密获取存储在 Cookie 中的用户名
    userName=CookieEncryptTool.decodeBase64(cookie.getValue());
    }
    if("COOKIE_PASSWORD".equals(cookie.getName())){
        //解密获取存储在 Cookie 中的密码
    password=CookieEncryptTool.decodeBase64(cookie.getValue());
    }
 }
}
%>
<center>
    <h2> 登录页面</h2>
    <form action ="LoginAction" method="post">
     <input type="hidden"    name="loginDate" value="<%=(new Date()).toLocaleString() %>"><br>
        用户名： <input type="text" maxlength="20" size="20" name="userName"><br><br>
        密       码： <input type="password" maxlength="20" size="20"
name="password"><br><br>
        <span>
                <input type="submit" value="登  录"/>
        <input checked="checked" name="rememberMe" type="checkbox" value="true">记住密码
</span>
        </form>
</center>
```

```
</body>
</html>
```

示例 10-17：

```
package control;

import java.io.IOException;
import java.text.DateFormat;
import java.text.ParseException;
import java.text.SimpleDateFormat;
import java.util.ArrayList;
import java.util.Date;
import java.util.List;

import javax.servlet.ServletContext;
import javax.servlet.ServletException;
import javax.servlet.annotation.WebServlet;
import javax.servlet.http.Cookie;
import javax.servlet.http.HttpServlet;
import javax.servlet.http.HttpServletRequest;
import javax.servlet.http.HttpServletResponse;

import dao.UserDao;
import domain.User;
import util.CookieEncryptTool;

/**
 * 登录功能实现
 */
@WebServlet("/LoginServlet")
public class LoginServlet extends HttpServlet {
  private static final long serialVersionUID = 1L;

  public LoginServlet() {
       super();
  }

  public void doGet(HttpServletRequest request, HttpServletResponse response) throws ServletException,
IOException {
       doPost(request, response);
  }

  public void doPost(HttpServletRequest request, HttpServletResponse response) throws ServletException,
IOException {
       // 设置请求和相应编码
       request.setCharacterEncoding("utf-8");
```

```
response.setContentType("text/html;charset=utf-8");
// 获取请求参数
String userName = request.getParameter("userName");
String password = request.getParameter("password");
String rememberMe = request.getParameter("rememberMe");
String loginDate = request.getParameter("loginDate");
System.out.println("loginDate" + loginDate);
DateFormat format = new SimpleDateFormat("yyyy-MM-dd HH:mm:ss");
Date date = null;
try {
    date = format.parse(loginDate);
} catch (ParseException e) {
    e.printStackTrace();
}

int uid = 0;
User user = new User(userName, password);
// 登录验证
UserDao dao = new UserDao();
uid = dao.userLogin(user);
if (uid == 0) {
    response.sendRedirect("login.jsp");
    return;
}
// 用户登录成功，将用户信息存入 session
user.setId(uid);
user.setLoginDate(date);
// 更新用户的登录时间
dao.updateTime(user);
request.getSession().setAttribute("userName", userName + uid);
request.getSession().setAttribute("SESSION_USER", user);
// 通过 Cookie 记住用户名和密码
rememberMe(rememberMe, userName, password, request, response);

ServletContext ctx = this.getServletContext();
List userList = (List) ctx.getAttribute("userList");
if (userList == null) {// 没有用户登录
    userList = new ArrayList();
    userList.add(userName + uid);
    ctx.setAttribute("userList", userList);
} else { // 已经有用户登录
    if (userList.contains(userName + uid) == false) {
        userList.add(userName + uid);
    }
}
response.sendRedirect("mainPage.jsp");
```

```
    }

    private void rememberMe(String rememberMe, String userName, String password, HttpServletRequest
request,
            HttpServletResponse response) {
        // 判断是否需要通过 Cookie 记住用户名和密码
        if ("true".equals(rememberMe)) {
            // 记住邮箱及密码
            Cookie cookie = new Cookie("COOKIE_USERNAME",
CookieEncryptTool.encodeBase64(userName));
            cookie.setPath("/");
            cookie.setMaxAge(365 * 24 * 3600);
            response.addCookie(cookie);
            cookie = new Cookie("COOKIE_PASSWORD",
CookieEncryptTool.encodeBase64(password));
            cookie.setPath("/");
            cookie.setMaxAge(365 * 24 * 3600);
            response.addCookie(cookie);
        } else {
            // 将用户名及密码 Cookie 删除
            Cookie[] cookies = request.getCookies();
            if (cookies != null) {
                for (Cookie cookie : cookies) {
                    if ("COOKIE_USERNAME".equals(cookie.getName()) ||
"COOKIE_PASSWORD".equals(cookie.getName())) {
                        cookie.setMaxAge(0);
                        cookie.setPath("/");
                        response.addCookie(cookie);
                    }
                }
            }
        }

    }
}
```

(6) 新建 mainPage.jsp 主页展示页面，以及 LogoutServlet 退出登录实现代码，如示例 10-18 和示例 10-19 所示。

示例 10-18：

```
<%@ page language="java" contentType="text/html; charset=UTF-8"
    pageEncoding="UTF-8"%>
    <%@page import="java.util.*" %>

<!DOCTYPE html>
<html>
<head>
```

```
<meta charset="UTF-8">
<title>玩家信息管理系统</title>
</head>
<body>
  <center>
      <h2>玩家信息管理系统</h2>
  </center>
  <hr/>
  <% int count=Integer.parseInt(application.getAttribute("access_count").toString());
          count++;
   %>
 当前访问次数：<%=count %>
 <%
     application.setAttribute("access_count",count);
  %>
<%--
在服务器启动之时，读取数据库存储的数据，
只要访问主页，就需要将 ServletContext 中的 access_count 做加 1 操作，再显示在页面上
退出 Web 应用时，将 access_count 写入到数据库表中  --%>
  <br>
    欢迎您，<%=session.getAttribute("userName")%>
    <br><br>
  当前玩家列表<br>
 <%
 List list=(List)application.getAttribute("userList");
 for (Object    o:list){
      out.println((String)o);
      out.println("<br/>");
 }
  System.out.println(session.getAttribute("userName"));
  %>
  <hr/>
  <a href="LogoutAction" >退出</a>
  </body>

</html>
```

示例 10-19：

```
package control;

import java.io.IOException;
import java.io.PrintWriter;
import java.util.List;

import javax.servlet.ServletContext;
import javax.servlet.ServletException;
```

```
import javax.servlet.annotation.WebServlet;
import javax.servlet.http.HttpServlet;
import javax.servlet.http.HttpServletRequest;
import javax.servlet.http.HttpServletResponse;
import javax.servlet.http.HttpSession;

import domain.User;

/**
 * 退出系统
 */
@WebServlet("/LogoutServlet")
public class LogoutServlet extends HttpServlet {
 private static final long serialVersionUID = 1L;

 public LogoutServlet() {
      super();
 }

 protected void doGet(HttpServletRequest request, HttpServletResponse response)
            throws ServletException, IOException {
     ServletContext ctx = this.getServletContext();
     // 退出系统后，从当前的 ServletContext 中删除用户信息，并切换到登录界面
     HttpSession session = request.getSession();
     List list = (List) ctx.getAttribute("userList");
     if (ctx != null) {
          User user = (User) session.getAttribute("SESSION_USER");
          list.remove(user.getUserName() + user.getId());
     }
     session.invalidate();

     response.setContentType("text/html");
     response.setCharacterEncoding("utf-8");
     PrintWriter out = response.getWriter();
     out.println("<!DOCTYPE HTML PUBLIC \"-//W3C//DTD HTML 4.01 Transitional//EN\">");
     out.println("<HTML>");
     out.println("   <HEAD><TITLE>退出登录</TITLE></HEAD>");
     out.println("   <BODY>");
     out.print("退出系统");
     out.println("   </BODY>");
     out.println("</HTML>");
     out.flush();
     out.close();

 }

 protected void doPost(HttpServletRequest request, HttpServletResponse response)
```

```
                throws ServletException, IOException {
        doGet(request, response);
    }

}
```

(7) 编写两个工具类，一个是 DBHelper，用于对数据库连接进行处理，一个是 CookieEncrytTool 类，用于对 Cookie 进行加密和解密，代码如示例 10-20 和示例 10-21 所示。

示例 10-20：

```
package util;

import java.io.IOException;
import java.io.InputStream;
import java.sql.Connection;
import java.sql.DriverManager;
import java.util.Properties;

public class DBHelper {
    private static String driver;
    private static String url;
    private static String username;
    private static String password;
    private static Connection conn;

    /**
     * 初始化各种 JDBC 数据库连接参数
     */
    private static void init() {
        try {
            // 创建属性集对象
            Properties props = new Properties();
            InputStream fis = DBHelper.class.getResourceAsStream("/jdbc.properties");
            props.load(fis);
            driver = props.getProperty("driver");
            url = props.getProperty("url");
            username = props.getProperty("username");
            password = props.getProperty("password");
        } catch (IOException e) {
            e.printStackTrace();
        }
    }

    /**
     * 提供连接
     *
```

```
     * @return 连接
     */
    public static Connection getConnection() {
        init();
        try {
            Class.forName(driver);
            conn = DriverManager.getConnection(url, username, password);
        } catch (Exception e) {
            e.printStackTrace();
        }
        return conn;
    }

    /**
     * 关闭连接
     */
    public static void closeConn() {
        try {
            if (conn != null) {
                conn.close();
                conn = null;
            }
        } catch (Exception e) {
            e.printStackTrace();
        }
    }
}
```

示例 10-21：

```
package util;

import org.apache.tomcat.util.codec.binary.Base64;

public class CookieEncryptTool {
    /**
     * Base64 加密
     */
    public static String encodeBase64(String cleartext) {
        try {
            cleartext = new String(Base64.encodeBase64(cleartext.getBytes("utf-8")));
        } catch (Exception e) {
            e.printStackTrace();
        }
        return cleartext;
    }
    /**
```

```
    * Base64 解密
    */
    public static String decodeBase64(String ciphertext) {
        try {
            ciphertext = new String(Base64.decodeBase64(ciphertext.getBytes("utf-8")));
        } catch (Exception e) {
            e.printStackTrace();
        }
        return ciphertext;
    }
}
```

(8) 在 src 目录下新建一个 jdbc.properties 文件，并且在 web.xml 中配置好 Servlet 文件，代码如示例 10-22 所示。

示例 10-22：

```
<?xml version="1.0" encoding="UTF-8"?>
<web-app xmlns:xsi="http://www.w3.org/2001/XMLSchema-instance"
 xmlns="http://xmlns.jcp.org/xml/ns/javaee"
 xsi:schemaLocation="http://xmlns.jcp.org/xml/ns/javaee
http://xmlns.jcp.org/xml/ns/javaee/web-app_3_1.xsd"
 id="WebApp_ID" version="3.1">
 <display-name>ChapterTenListener</display-name>
 <servlet>
     <servlet-name>LogoutServlet</servlet-name>
     <servlet-class>control.LogoutServlet</servlet-class>
 </servlet>
 <servlet>
     <servlet-name>LoginServlet</servlet-name>
     <servlet-class>control.LoginServlet</servlet-class>
 </servlet>
 <servlet-mapping>
     <servlet-name>LoginServlet</servlet-name>
     <url-pattern>/LoginAction</url-pattern>
 </servlet-mapping>
 <servlet-mapping>
     <servlet-name>LogoutServlet</servlet-name>
     <url-pattern>/LogoutAction</url-pattern>
 </servlet-mapping>
 <welcome-file-list>
     <welcome-file>login.jsp</welcome-file>
 </welcome-file-list>
</web-app>
```

(9) 在此启动服务器，在浏览器中输入地址 http://localhost:8080/ChapterTenListener/，进入登录页面，同时打开多个不同的浏览器，分别输入数据库中事先准备的账号和密码进行登录。然后再关闭服务器，此时会发现数据库中会保存当前访问的次数。再次登录进行以上操作，得到的页面信息如图 10-8 所示。

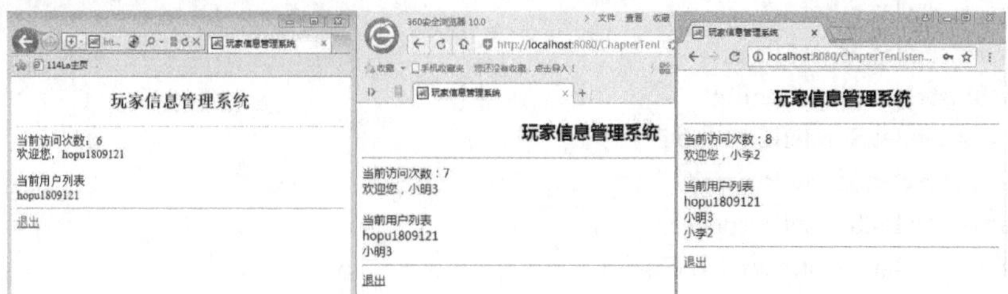

图 10-8

我们将其中一个浏览器用户退出，再刷新另外两个浏览器，此时页面信息如图 10-9 所示。

图 10-9

◆ 第二阶段 ◆

练习 2：监听器和过滤器综合过滤未登录用户访问主页，并限制用户在线时长

【问题描述】

在上面一个案例中，进一步改进，过滤未登录的用户访问主页，并且，控制玩家在线时长不得超过 8 小时，否则就强制下线，需要重新登录。

【问题分析】

创建一个过滤器，拦截那些处于未登录状态的用户，将这些用户的请求地址转发到登录页面；再创建一个监听器，监听用户的请求，当用户有请求时，读取他的登录时间和当前时间进行对比，超过 8 小时就将 session 销毁。用户再次刷新页面就会重写调转到登录页。

【参考步骤】

(1) 在上一工程项目 ChapterTenListener 下，再新建一个 AuthorityFilter 过滤器，代码如示例 10-23 所示。

示例 10-23：

```
package filter;

import java.io.IOException;
```

```
import javax.servlet.Filter;
import javax.servlet.FilterChain;
import javax.servlet.FilterConfig;
import javax.servlet.ServletException;
import javax.servlet.ServletRequest;
import javax.servlet.ServletResponse;
import javax.servlet.annotation.WebFilter;
import javax.servlet.http.HttpServletRequest;
import javax.servlet.http.HttpServletResponse;
import javax.servlet.http.HttpSession;

/**
 * 玩家访问权限过滤器
 *
 * @author zl
 */
@WebFilter("/AuthorityFilter")
public class AuthorityFilter implements Filter {
 private String loginPage = "login.jsp";

 public AuthorityFilter() {
 }

 public void destroy() {
      this.loginPage = null;
 }

 public void doFilter(ServletRequest request, ServletResponse response, FilterChain chain)
          throws IOException, ServletException {
      HttpServletRequest req = (HttpServletRequest) request;
      HttpServletResponse resp = (HttpServletResponse) response;
      HttpSession session = req.getSession();
      // 判断被拦截的请求用户是否处于登录状态
      if (session.getAttribute("SESSION_USER") == null) {
           // 获取被拦截的请求地址参数
           String requestURI = req.getRequestURI();
           if (req.getQueryString() != null) {
                requestURI += "?" + req.getQueryString();
           }
           // 将请求地址保存到 request 对象中转发到登录页面
           req.setAttribute("requestURI", requestURI);
           request.getRequestDispatcher("/" + loginPage).forward(request, response);
           return;
      } else {

           chain.doFilter(request, response);
```

```
        }
    }

    public void init(FilterConfig fConfig) throws ServletException {
        // 当请求被拦截时获取转向的页面
        loginPage = fConfig.getInitParameter("loginPage");
        if (null == loginPage) {
            loginPage = "login.jsp";
        }
    }

}
```

(2) 再新建一个 UserRequestListener 监听器，用户监考用户在线时长，如示例 10-24 所示。

示例 10-24：

```
package listener;

import java.util.Date;
import java.util.List;

import javax.servlet.ServletRequestEvent;
import javax.servlet.ServletRequestListener;
import javax.servlet.annotation.WebListener;
import javax.servlet.http.HttpServletRequest;

import dao.UserDao;
import domain.User;

/**
 * 监听用户在线时长
 *
 */
@WebListener
public class UserRequestListener implements ServletRequestListener {

    public void requestDestroyed(ServletRequestEvent sre) {
    }

    public void requestInitialized(ServletRequestEvent sre) {
        // TODO Auto-generated method stub
        HttpServletRequest request = (HttpServletRequest) sre.getServletRequest();
        String requestURI = request.getRequestURI();
        System.out.println("requestURI:" + requestURI);
        User user = (User) request.getSession().getAttribute("SESSION_USER");
        Date date = new Date();
```

```
        UserDao dao = new UserDao();
        if (user != null) {
            Date loginDate = dao.getLoginDate(user.getId());
            System.out.println("loginDate:" + loginDate);
            System.out.println("date:" + date);
            Long onlineTime = date.getTime() - loginDate.getTime();
            System.out.println("onlineTime:" + onlineTime);
            if (onlineTime / (1000 * 60 * 60) >= 8) {
                System.out.println("在线时长超时，你已被迫下线");
                List list = (List) sre.getServletContext().getAttribute("userList");
                list.remove(user.getUserName() + user.getId());
                request.getSession().invalidate();

            }
        }
    }

}
```

(3) 修改 web.xml，代码如示例 10-25 所示。

示例 10-25：

```
<?xml version="1.0" encoding="UTF-8"?>
<web-app xmlns:xsi="http://www.w3.org/2001/XMLSchema-instance"
xmlns="http://xmlns.jcp.org/xml/ns/javaee" xsi:schemaLocation="http://xmlns.jcp.org/xml/ns/javaee
http://xmlns.jcp.org/xml/ns/javaee/web-app_3_1.xsd" id="WebApp_ID" version="3.1">
    <display-name>ChapterTenListener</display-name>
     <filter>
    <filter-name>AuthorityFilter</filter-name>
    <filter-class>filter.AuthorityFilter</filter-class>
    </filter>
    <filter-mapping>
   <filter-name>AuthorityFilter</filter-name>
   <url-pattern>/mainPage.jsp</url-pattern>
    </filter-mapping>
    <servlet>
       <servlet-name>LogoutServlet</servlet-name>
       <servlet-class>control.LogoutServlet</servlet-class>
    </servlet>
    <servlet>
       <servlet-name>LoginServlet</servlet-name>
       <servlet-class>control.LoginServlet</servlet-class>
    </servlet>
    <servlet-mapping>
       <servlet-name>LoginServlet</servlet-name>
       <url-pattern>/LoginAction</url-pattern>
    </servlet-mapping>
```

```
    <servlet-mapping>
        <servlet-name>LogoutServlet</servlet-name>
        <url-pattern>/LogoutAction</url-pattern>
    </servlet-mapping>
    <welcome-file-list>
        <welcome-file>login.jsp</welcome-file>
    </welcome-file-list>
</web-app>
```

(4) 启动服务器，当我们直接访问 http://localhost:8080/ChapterTenListener/mainPage.jsp 页面时，会自动调转到登录页面进行登录操作，登录成功后，为了便于测试，同学们可以将在线时长控制在 1 分钟。在登录成功一分钟后，再次刷新页面，就会自动退出主页，调转到登录页面。这就是我们上一章节学习的过滤器和本章节学习的监听器的综合案例。

【拓展作业】

1. 简要描述过滤器和监听器的功能和区别。

2. 编写一个监听器，通过监听器记录在线用户的姓名，在页面显示用户的姓名，同时实现对某个用户的强制下线功能。

单元十一 MVC 模式

课程目标

- ▶ 掌握 Model 1 体系结构
- ▶ 掌握 Model 2 体系结构
- ▶ 掌握 MVC 应用程序

简介

MVC 模式是 Model-View-Controller 的缩写，中文翻译为“模型-视图-控制器”，最早是由 SmallTalk 社区开发并提出的，应用于 GUI 用户交互应用程序中。MVC 模式试图将应用程序分成 3 部分，并且详细定义这 3 部分之间的交互，从而降低它们之间的耦合度，让每一部分都专注于自己的职责。用户和视图(View)进行交互，输入数据并单击按钮，控制器(Controller)收到来自视图的事件，并对模型(Model)进行操作，根据用户提供的数据更新模型(Model)，视图同时也会收到“模型更新”的事件通知，从而自我更新，将结果呈现给用户。

SmallTalk 语言和 Java 语言有很多相似之处，都是面向对象的程序设计语言，很自然的，SUN 在 Java EE 应用程序中就推荐 MVC 模式作为开发 Web 应用的架构模式。

11.1 MVC 模式在 Web 中的应用

现在的 Web 应用程序大多采用 MVC 的模式设计。下面介绍 MVC 3 个部分在 Java EE 架构中处于什么位置，这样有助于理解 MVC 模式的实现。MVC 与 Java EE 架构的对应关系是：View 处于 Web 层或者说是 Client 层，通常是 JSP 或 Servlet，即页面显示部分。Controller 也处于 Web 层，通常用 Servlet 来实现，即页面显示的逻辑部分实现。Model 处于 Middle 层，通常用服务端的 JavaBean 或者 EJB 实现，即业务逻辑部分的实现。

MVC 模式的核心思想是把一个应用的输入、处理、输出流程按照 Model、View、Controller 的方式进行分离，这样一个应用被分成 3 个层：模型层、视图层、控制层。

视图(View)代表用户交互界面，对于 Web 应用来说，可以概括为 HTML、JSP 界面，更有可能为 XHTML、XML 和 Applet 界面。随着应用的复杂性和规模性增加，界面的处理也变得具有挑战性。一个应用可能有很多不同的视图，MVC 设计模式中对于视图的处理仅限于视图上数据的采集和处理，以及用户的请求，而不包括在视图上的业务流程的处理。业务流程的处理交予模型(Model)处理。例如，一个订单的视图只接受来自模型的数据并显示给用户，以及将用户界面的输入数据和请求传递给控制器和模型。

模型(Model)就是业务流程和状态的处理以及业务规则的制定。业务流程的处理过程对其他层来说是暗箱操作，模型接收视图请求的数据，并返回最终的处理结果。业务模型的设计可以说是 MVC 最主要的核心。例如，Java EE 框架中的 EJB 模型就是一个典型，它从应用技术实现的角度对模型做了进一步的划分，以便充分利用现有的组件，但它不能作为应用设计模型的框架。它仅仅告诉用户按这种模型设计就可以利用某些技术组件，从而减少了技术上的困难。对一个开发者来说，就可以专注于业务模型的设计。MVC 设计模式说明，把应用的模型按一定的规则抽取出来，抽取的层次很重要，这也是判断开发人员的设计是否优秀的依据。抽象与具体不能隔得太远，也不能太近。MVC 并没有提供模型的设计方法，而只告诉用户应该组织管理这些模型，以便于模型的重构和提高重用性。

控制器(Controller)可以理解为从用户接收请求，将模型与视图匹配在一起，共同完成用户的请求。划分控制层的作用也很明显，它清楚地告诉用户，它就是一个分发器，选择什么样的模型、什么样的视图，可以完成什么样的用户请求。控制层并不做任何数据处理。例如，用户单击一个链接，控制层接受请求后，并不处理业务信息，它只把用户的信息传递给模型，告诉模型做什么，选择符合要求的视图返回给用户。因此，一个模型可能对应多个视图，一个视图可能对应多个模型。

模型、视图与控制器的分离，使得一个模型可以具有多个显示视图。如果用户通过某个视图的控制器改变了模型的数据，所有其他依赖于这些数据的视图都应反映出这些变化。因此，无论何时发生了何种数据变化，控制器都会将变化通知所有的视图，使显示更新。

11.2　JSP Model 1体系结构

在 Model 1 模式下，整个 Web 应用几乎全部由 JSP 页面组成，JSP 页面接受处理客户端请求，对请求处理后直接做出响应。用少量的 JavaBean 来处理数据库连接、数据库访问等操作。Model 1 体系结构如图 11-1 所示。

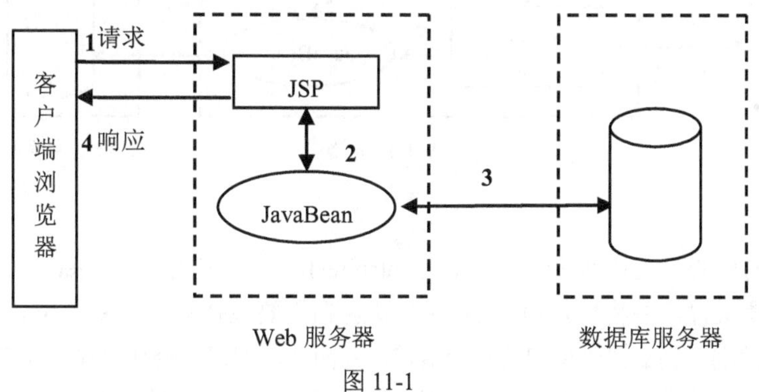

图 11-1

Model 1 模式的实现比较简单，适用于快速开发小规模项目。但从工程化的角度看，它的局限性非常明显：JSP 页面身兼 View 和 Controller 两种角色，将控制逻辑和表现逻辑混杂在一起，JSP 页面的编写者必须既是网页的设计者又是 Java 的开发者。实际上，多数的 Web 开发人员只精通网页的设计，能够设计出精美的页面，但代码设计很糟糕。而精通 Java 代码的人员，页面设计得很难看。兼具两种能力的人员很少。

- 内嵌的逻辑使开发人员要理解应用程序的整个流程，必须浏览所有的页面，工作量很大。HTML 标签、Java 代码和 JavaScript 代码集中在同一个页面中，使调试变得很困难。
- 强耦合性，变更业务逻辑和数据会涉及多个页面。

为解决上面的问题，SUN 公司制定了 JSP Model 2 规范。

11.3 JSP Model 2 体系结构

JSP Model 2 结构采用 MVC 模式，其主要思想是：JSP 只是用来接收用户的输入和把输出结果以一定的格式返回给客户，是 MVC 的视图部分。客户的请求一般由 Servlet 来处理，Servlet 还可以通过访问服务器端的组件来完成事务处理，是 MVC 模式的控制器部分。Servlet 产生的结果或者是传递给 JSP 页面，或者是使 JSP 能够访问这些结果，然后再由 JSP 把这些结果呈现给客户端。

Model 2 的主要优点是使 JSP 页面更易于理解和维护，目标是把事务逻辑和实现层清楚地分开。这是突破性的软件设计方法，它清晰地分离了表达和内容，明确了角色定义及开发者和网页设计者的分工。实际上，项目越复杂，Model 2 的优势越明显。Model 2 的结构如图 11-2 所示。

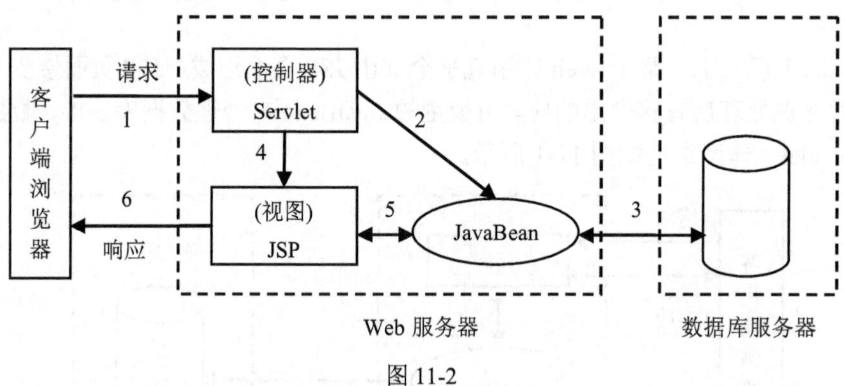

图 11-2

Model 2 的核心是前端控制器(Front Controller)，它被实现为 Servlet，其主要作用是初始化并解析配置文件，并将用户请求的 URL 映射到需要被执行的命令实例，命令实例其实就是 Action。前端控制器接受用户的每个请求，并简单地把请求委托给调度器(Dispatcher)，由调度器执行相应的动作(Action)。调度器把 Action 返回的 URL 返回给前端控制器，并由前端控制器负责转发。

11.3.1 实现 MVC 框架

下面来实现一个简单的 MVC 框架，并从中体会 MVC 框架的思想，然后复用这个框架来完成 JSP Model 2 结构的网上商店。该 MVC 的框架如图 11-3 所示。

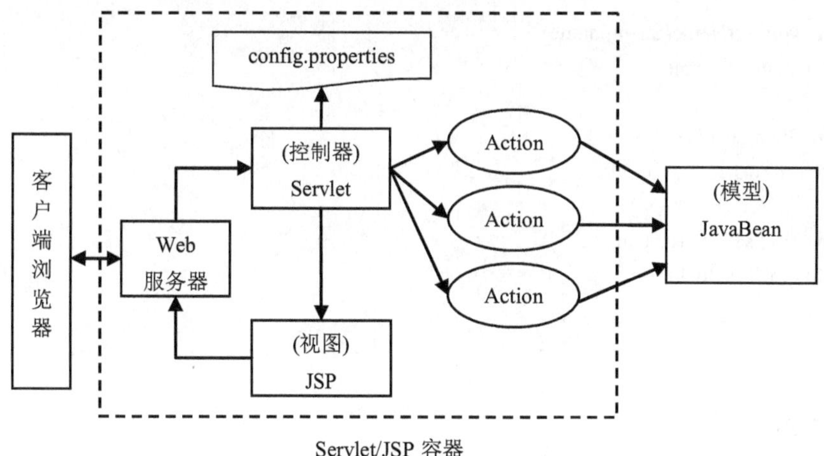

图 11-3

一般来说，推出框架的终极目标是要实现代码的复用，不同的应用建立在同一个框架的基础之上，不然就失去了框架的意义。要实现如图 11-3 所示的 MVC 框架，分析如下。

(1) 需要建立属性文件，用来存放不同应用中要用到的页面的名称、处理业务要用到的类的名称，本例中采用键值对的形式。属性文件的例子代码如示例 11-1 所示，这个文件必须是以.properties 为扩展名，存放在工程的 WEB-INF/目录下面。

示例 11-1：

```
doAdd=/doAdd,model.AddToCartBiz,add,index.jsp
doDelete=/doDelete,model.DeleteBiz,delete,index.jsp
doCheckOut=/doCheckOut,model.CheckOutBiz,checkOut,checkOut.jsp
```

(2) 需要建立两个 JavaBean 模型，分别是 ActionForward 和 ActionMapping。其中，ActionForward 封装了转向操作所需要信息的一个模型，包括 name 和转向 toUrl，name 属性用来指定要转向页面的逻辑名，toUrl 用来指定物理的页面名。ActionMapping 类将每一个用户请求的 Action 封装成一个 ActionMapping 对象，把所有 ActionMapping 对象构成一个 Map，可以根据键来查找对应的 ActionMapping 对象，并存储在 ServletContext 中，供整个框架使用。ActionForward 和 ActionMapping 类的代码分别如示例 11-2 和示例 11-3 所示。

示例 11-2：

```
package model;

public class ActionForward {
    private String name; //要转向的页面逻辑名称
    private String toUrl;//要转向的页面实际名称
    public String getName() {
        return name;
    }
```

```
        public void setName(String name) {
            this.name = name;
        }
        public String getToUrl() {
            return toUrl;
        }
        public void setToUrl(String toUrl) {
            this.toUrl = toUrl;
        }
}
```

示例 11-3：

```
package model;
import java.util.Map;

public class ActionMapping {
        private String path; //请求的 URL
        private String className;//处理请求的类名
        private Map<String, ActionForward> forwards;
        public String getPath() {
            return path;
        }
        public void setPath(String path) {
            this.path = path;
        }
        public String getClassName() {
            return className;
        }
        public void setClassName(String className) {
            this.className = className;
        }
        public Map<String, ActionForward> getForwards() {
            return forwards;
        }
        public void setForwards(Map<String, ActionForward> forwards) {
            this.forwards = forwards;
        }
}
```

(3) 需要建立一个工具类来解析 Properties 文件。例如 PropertyUtil 类，代码如示例 11-4 所示。这里需要注意的是，该类的 parse()方法返回一个 ActionMapping 对象的 Map 集合，入参是给定文件的路径和 Servlet 的上下文。

示例 11-4：

```
package util;
```

```
import java.io.IOException;
import java.io.InputStream;
import java.util.Properties;
import java.util.Map;
import java.util.HashMap;
import java.util.Enumeration;
import javax.servlet.ServletContext;
import model.*;

public class PropertyUtil {
    private static PropertyUtil _self = null;
    private static Properties p = new Properties();

    public static synchronized PropertyUtil getInstance() throws IOException {
        if (_self == null) {
            _self = new PropertyUtil();
        }
        return _self;
    }

    public Map<String, ActionMapping> parse(String file_path,
            ServletContext context) throws Exception {
        InputStream is = context.getResourceAsStream(file_path);
        p.load(is);
        return parseFile(p);
    }
     //解析配置文件 config.properties
private Map<String, ActionMapping> parseFile(Properties prop)
            throws Exception {
        //生成 Mapping Map 类
    Map<String, ActionMapping> actions = new HashMap<String, ActionMapping>();
        //生成 Forward Map 类
    Map<String, ActionForward> forwards = new HashMap<String, ActionForward>();
        //返回所有键值的枚举
    Enumeration enu = prop.propertyNames();
        while (enu.hasMoreElements()) {
            //获取单个键值
            String key = (String) enu.nextElement();
            //根据键值获取字符串
            String value = (String) prop.get(key);
            //使用","分隔字符串
            String[] values = value.split(",");
            //生成 ActionForward 对象
            ActionForward forward = new ActionForward();
            forward.setName(values[2]);//逻辑名
            forward.setToUrl(values[3]);//物理名
```

```
                forwards.put(values[2], forward);
                //生成 ActionMapping 对象
                ActionMapping actionMapping = new ActionMapping();
                actionMapping.setPath(values[0]);//请求的 URL
                actionMapping.setClassName(values[1]);//处理请求对应的类
                actionMapping.setForwards(forwards);
                actions.put(key, actionMapping);
            }
            return actions;
        }
    }
```

(4) 定义一个 Action 接口，它是一个命令接口，每一个实现了此接口的 Action 都封装了某一个请求，并在重写的 execute()方法中实现不同的业务操作。Execute()方法有两个入参，分别是 HttpServletRequest 和 ServletContext 类型，返回一个 ActionForward 提供给 Dispatcher 调用。代码如示例 11-5 所示。

示例 11-5：

```
package model;

import javax.servlet.http.HttpServletRequest;
import javax.servlet. ServletContext;

public interface Action {
    public ActionForward execute(HttpServletRequest request,ServletContext context);
}
```

(5) 建立一个调度器类 ServletDispatcher，负责调用 Action 去执行业务逻辑。调度器服务于前端控制器，它把对模型的更新委托给 Action，又提供页面选择给前端控制器。代码如示例 11-6 所示。

示例 11-6：

```
package model;

import javax.servlet.ServletContext;
import javax.servlet.http.HttpServletRequest;
import java.util.Map;

public class ServletDispatcher {

    public String getNextPage(HttpServletRequest request,ServletContext context) {
        Map<String, ActionMapping> actions = (Map<String, ActionMapping>)
  context.getAttribute("Actions");
        String reqPath =(String)request.getAttribute("RequestPath");
        ActionMapping actionMapping = actions.get(reqPath);
```

```
        String forward_name = "";
        ActionForward actionForward;
        try {
            // 根据请求 URL 生成 Action 对象
            Class c = Class.forName(actionMapping.getClassName());
            Action action = (Action) c.newInstance();
            //执行 execute 方法
            actionForward = action.execute(request, context);
            forward_name = actionForward.getName();
        } catch (Exception e) {
            e.printStackTrace();
        }
        actionForward =actionMapping.getForwards().get(forward_name);
        if (actionForward == null) {
            return null;
        } else
            return actionForward.getToUrl();
    }
}
```

(6) 建立前端控制器(FrontController)，它提供了一个统一的位置来封装公共请求，主要作用是初始化并解析配置文件，接受每个请求，并简单地把请求委托给调度器(Dispatcher)，由调度器执行相应的动作(Action)。调度器把 Action 返回的 URL 返回给前端控制器，前端控制器负责页面的转发。代码如示例 11-7 所示。

示例 11-7：

```
package control;
import java.io.IOException;
import javax.servlet.RequestDispatcher;
import javax.servlet.ServletException;
import javax.servlet.http.HttpServlet;
import javax.servlet.http.HttpServletRequest;
import javax.servlet.http.HttpServletResponse;
import javax.servlet.ServletContext;
import java.util.Map;
import model.*;
import util.PropertyUtil;

public class FrontControllerServlet extends HttpServlet {
    private ServletContext context;

    // 初始化前端控制器
    public void init() throws ServletException {
        context = getServletContext();
        String config_file =getServletConfig().getInitParameter("config");
        String dispatcher_name =getServletConfig().getInitParameter(
```

```
                "defaultDispatcher");
        try {
            Map<String, ActionMapping> resources = PropertyUtil.getInstance().parse(config_file,
                    context);
            context.setAttribute("Actions", resources); // 存储在 ServletContext 中
            Class c = Class.forName(dispatcher_name);
            ServletDispatcher dispatcher = (ServletDispatcher) c.newInstance();
            context.setAttribute("Dispatcher", dispatcher); // 放在 ServletContext
        } catch (Exception e) {
            e.printStackTrace();
        }
    }

    public void doGet(HttpServletRequest request, HttpServletResponse response)
            throws ServletException, IOException {
        request.setCharacterEncoding("utf-8");
        response.setContentType("text/html;charset=utf-8");
        this.process(request, response);
    }

    public void doPost(HttpServletRequest request, HttpServletResponse response)
            throws ServletException, IOException {
        request.setCharacterEncoding("utf-8");
        response.setContentType("text/html;charset=utf-8");
        this.process(request, response);
    }

    protected void process(HttpServletRequest request,
            HttpServletResponse response) throws ServletException, IOException {
        // 获取客户端请求的全 URL
        String reqURI = request.getRequestURI();
        int i = reqURI.lastIndexOf(".");
        String contextPath = request.getContextPath();
        // 获取 action 的 URL
        String path = reqURI.substring(contextPath.length() + 1, i);
        request.setAttribute("RequestPath", path);
        // 获取默认的分发器对象
        ServletDispatcher   dispatcher = (ServletDispatcher) context
                .getAttribute("Dispatcher");
        // 调用 Dispatcher 的 getNextPage
        String nextPage = dispatcher.getNextPage(request, context);
        RequestDispatcher forwarder =
            request.getRequestDispatcher("/" + nextPage);
        forwarder.forward(request, response); // 转发页面
    }
}
```

(7) 在容器启动时加载前端控制器，并且需要给这个前端控制器添加初始化参数。web.xml 的代码如示例 11-8 所示，注意 Servlet 的 url-pattern。

示例 11-8：

```
<?xml version="1.0" encoding="UTF-8"?>
<web-app version="2.5" xmlns="http://java.sun.com/xml/ns/javaee"
    xmlns:xsi="http://www.w3.org/2001/XMLSchema-instance"
    xsi:schemaLocation="http://java.sun.com/xml/ns/javaee
    http://java.sun.com/xml/ns/javaee/web-app_2_5.xsd">
    <servlet>
        <servlet-name>FrontControllerServlet</servlet-name>
        <servlet-class>control.FrontControllerServlet</servlet-class>
        <init-param>
            <param-name>config</param-name>
            <param-value>/WEB-INF/config.properties</param-value>
        </init-param>
        <init-param>
        <param-name>defaultDispatcher</param-name>
        <param-value>model.ServletDispatcher</param-value>
        </init-param>
        <load-on-startup>0</load-on-startup>
    </servlet>
    <servlet-mapping>
        <servlet-name>FrontControllerServlet</servlet-name>
        <url-pattern>*.action</url-pattern>
    </servlet-mapping>
    <welcome-file-list>
        <welcome-file>index.jsp</welcome-file>
    </welcome-file-list>
</web-app>
```

11.3.2　使用 MVC 框架

如前所述，框架的目标就是要实现代码的复用，针对不同的应用，要做的事情就是实现视图以及其他要完成的具体业务，框架提供的是通用的代码，可以直接拿来使用。下面来实现简单的网上购物的应用。

网上购物页面有 3 个，作为视图部分，分别是 index.jsp、shoppingCart.jsp 和 checkOut.jsp，其中，index.jsp 页面用下拉列表的方式显示商品，shoppingCart.jsp 页面用来显示当前购物车里的商品列表，checkOut.jsp 页面用来显示生成的订单信息。

(1) 业务分 3 种，作为模型部分，分别是把商品添加到购物车、从购物车中删除商品和生成订单。3 种业务使用 3 个JavaBean来实现，分别是AddToCartAction、DeleteAction和CheckOutAction，它们都实现了Action接口，并重写了接口中定义的execute()方法来处理上面的业务。

(2) 站点的商品是 CD，使用 CD JavaBean 来实现。

(3) 3 个页面的代码分别如示例 11-9、示例 11-10 和示例 11-11 所示。

示例 11-9：

```
<%@ page contentType="text/html; charset=utf-8"%>

<html>
  <head><title>非常动听音乐</title></head>
<body bgcolor="#33CCFF">
<font face="隶书" size="+3">非常动听音乐吧</font>
<hr>
<form name="shoppingForm" action="doAdd.action" method="post">
请选择你喜爱的 CD:
<select name="CD">
<option>Yuan | The Guo Brothers | China | $14.95</option>
<option>Drums of Passion | Babatunde Olatunji | Nigeria | $16.95</option>
<option>Kaira | Tounami Diabate| Mali | $16.95</option>
<option>The Lion is Loose | Eliades Ochoa | Cuba | $13.95</option>
<option>Dance the Devil Away | Outback | Australia | $14.95</option>
<option>Record of Changes | Samulnori | Korea | $12.95</option>
<option>Djelika | Tounami Diabate | Mali | $14.95</option>
<option>Rapture | Nusrat Fateh Ali Khan | Pakistan | $12.95</option>
<option>Cesaria Evora | Cesaria Evora | Cape Verde | $16.95</option>
<option>Ibuki | Kodo | Japan | $13.95</option>
</select>数量:
<input type="text" name="qty" SIZE="3" value="1">
<input type="hidden" name="action" value="ADD"><br>
<input type="submit" name="Submit" value="加入到购物车">
</form>
<jsp:include page="shoppingCart.jsp" flush="true" />
</body>
</html>
```

示例 11-10：

```
<%@ page contentType="text/html; charset=utf-8"%>
<%@ taglib prefix="c" uri="http://java.sun.com/jsp/jstl/core" %>
<html>
  <head><title>非常动听音乐－购物车</title></head>
<body bgcolor="#33CCFF">
<c:set var="buylist" value="${sessionScope.shoppingcart}"/>
<c:choose>
<c:when test="${empty buylist}">
  把你喜欢的 CD 添加到购物车
</c:when>
```

```
<c:otherwise>
<table border="0" cellpadding="0" width="100%" bgcolor="#FFFFFF">
<tr>
<td>分类</td><td>艺术家</td><td>国籍</td><td>价格</td><td>数量</td>
<td> </td>
</tr>
  <c:forEach var="anOrder" items="${buylist}" varStatus="index">
<tr>
<td>${anOrder.album}</td><td>${anOrder.artist}</td><td>${anOrder.country}</td>
<td>${anOrder.price}</td><td>${anOrder.quantity}</td>
<td>
<form name="deleteForm" action="doDelete.action" method="POST">
<input type="submit" value="删除">
<input type="hidden" name= "delindex" value='${index.count-1}'>
<input type="hidden" name="action" value="DELETE">
</form>
</td>
</tr>
</c:forEach>
</table>
<form name="checkoutForm" action="doCheckOut.action" method="POST">
<input type="hidden" name="action" value="CHECKOUT">
<input type="submit" name="Checkout" value="去收银台">
</form>
</c:otherwise>
</c:choose>
</body>
</html>
```

示例 11-11：

```
<%@ page contentType="text/html; charset=utf-8"%>
<%@ taglib prefix="c" uri="http://java.sun.com/jsp/jstl/core" %>
<html>
<head><title>非常动听音乐一收银台</title></head>
<body bgcolor="#33CCFF">
<font face="隶书" size="+3">非常动听音乐吧</font>
<hr>
<table border="0" cellpadding="0" width="100%" bgcolor="#FFFFFF">
<tr>
<td>分类</td><td>艺术家</td><td>国籍</td><td>价格</td><td>数量</td>
<td></td>
</tr>
<c:forEach var="acd" items="${sessionScope.shoppingcart}">
<tr>
<td><b>${acd.album}</b></td><td><b>${acd.artist}</b></td>
<td><b>${acd.country}</b></td><td><b>${acd.price}</b></td>
```

```
<td><b>${acd.quantity}</b></td>
</tr>
</c:forEach>
<tr>
<td></td><td></td><td><b>总价</b></td>
<td><b>${requestScope.amount }</b></td><td></td>
</tr>
</table>
<a href="index.jsp">回到主页</a>
</body>
</html>
```

CD JavaBean 代码如示例 11-12 所示。

示例 11-12：

```
package bean;
import java.io.Serializable;

public class CD implements Serializable {
   private String album;
   private String artist;
   private String country;
   private float price;
   private int quantity;
public String getAlbum() {
     return album;
}
public void setAlbum(String album) {
     this.album = album;
}
public String getArtist() {
     return artist;
}
public void setArtist(String artist) {
     this.artist = artist;
}
public String getCountry() {
     return country;
}
public void setCountry(String country) {
     this.country = country;
}
public float getPrice() {
     return price;
}
public void setPrice(float price) {
     this.price = price;
```

```
    }
    public int getQuantity() {
        return quantity;
    }
    public void setQuantity(int quantity) {
        this.quantity = quantity;
    }
}
```

AddToCartBiz 的代码如示例 11-13 所示。

示例 11-13：

```
package model;

import javax.servlet.http.HttpServletRequest;
import java.util.List;
import java.util.ArrayList;
import java.util.StringTokenizer;
import javax.servlet.ServletContext;
import bean.CD;
import javax.servlet.http.HttpSession;
public class AddToCartBiz implements Action {

    public ActionForward execute(HttpServletRequest request,
            ServletContext context) {
        String myCd = request.getParameter("CD");
        String qty = request.getParameter("qty");
        StringTokenizer t = new StringTokenizer(myCd, "|");
        String album = t.nextToken();
        String artist = t.nextToken();
        String country = t.nextToken();
        String price = t.nextToken();
        price = price.replace('$', ' ').trim();
        CD aCD = new CD();
        aCD.setAlbum(album);
        aCD.setArtist(artist);
        aCD.setCountry(country);
        aCD.setPrice((new Float(price)).floatValue());
        aCD.setQuantity((new Integer(qty)).intValue());
        boolean match = false;
        HttpSession session = request.getSession();
        List<CD> list =(List<CD>)session.getAttribute("shoppingcart");
        if (list == null) {
            // 将第一张 CD 放入购物车
            list = new ArrayList<CD>(); // 第一份订单
            list.add(aCD);
        } else {
```

```
            // 不是第一次购买
            for (int i = 0; i < list.size(); i++) {
                CD cd = (CD) list.get(i);
                if (cd.getAlbum().equals(aCD.getAlbum())) {
                    cd.setQuantity(cd.getQuantity()
                    + aCD.getQuantity());
                    list.set(i, cd);
                    match = true;
                } // if name matches 结束
            } // for 循环结束
            if (!match)
                list.add(aCD);
        }
        session.setAttribute("shoppingcart", list);
        ActionForward af = new ActionForward();
        af.setName("add");
        return af;
    }
}
```

CheckoutBiz 的代码如示例 11-14 所示。

示例 11-14：

```
package model;

import java.util.List;
import javax.servlet.http.HttpServletRequest;
import javax.servlet.ServletContext;
import javax.servlet.http.HttpSession;
import bean.CD;

public class CheckOutBiz implements Action {

    public ActionForward execute(HttpServletRequest request,
            ServletContext context) {
        HttpSession session = request.getSession();
        List<CD> list =
            (List<CD>) session.getAttribute("shoppingcart");
        float total = 0;
        for (int i = 0; i < list.size(); i++) {
            CD anOrder = (CD) list.get(i);
            float price = anOrder.getPrice();
            int qty = anOrder.getQuantity();
            total += (price * qty);
        }
        total += 0.005;
        String amount = new Float(total).toString();
```

```
        int n = amount.indexOf('.');
        amount = amount.substring(0, n + 3);
        request.setAttribute("amount", amount);
        ActionForward af = new ActionForward();
        af.setName("checkOut");
        return af;
    }
}
```

DeleteBiz 的代码如示例 11-15 所示。

示例 11-15：

```
package model;

import java.util.List;
import javax.servlet.http.HttpServletRequest;
import javax.servlet.ServletContext;
import javax.servlet.http.HttpSession ;
import bean.CD;

public class DeleteBiz implements Action    {
    public ActionForward execute(HttpServletRequest request,
    ServletContext context) {
        String del = request.getParameter("delindex");
        int index= (new Integer(del)).intValue();
        HttpSession session = request.getSession();
        List<CD> list =
            (List<CD>)session.getAttribute("shoppingcart");
        list.remove(index);
        ActionForward af = new ActionForward();
        af.setName("delete");
        return af;
    }
}
```

index.jsp 运行结果如图 11-4 所示。

图 11-4

从下拉列表框中选择 CD，加入到购物车，结果如图 11-5 所示。

图 11-5

用户可以选择从购物车中删除 CD 或是去收银台结账，单击“去收银台”按钮，结果如图 11-6 所示。

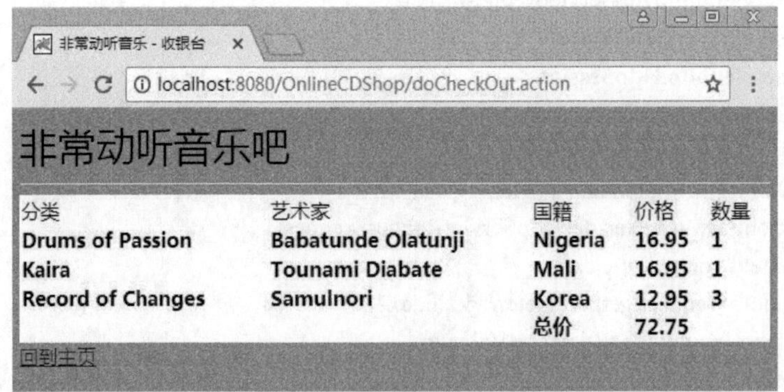

图 11-6

我们回头来看看上面的框架，其实它有个缺陷，就是每个 Action 完成之后，都转到某个物理页面，使用逻辑名来实现。想一想，在实际的业务中，业务处理后结果有多种，针对不同的情况，控制权就应该转到不同的结果页面上，这就需要在同一个 Action 下面，去配置多个逻辑页面名和物理页面名的映射。上面使用 Properties 配置文件是产生这个缺陷的根本原因。当然，可以重新设计配置文件和解析配置文件的 PropertyUtil 工具类，这个问题留给读者自己来完成。其实，学习的重点和目的是理解 MVC 这种模式的思想。关于 MVC 的模式，业界已经有了 MVC 模式的经典实现，那就是 Apache 组织的 Struts 框架，在后续的 Java 课程“Java 企业级开发框架——Struts”中会深入地学习 Struts 框架。这里，是想通过这个简单的框架自然地过渡到 Struts 框架，在 Struts 的实现里，配置文件变成了 struts-config.xml，解决了上面一对多的问题。

【单元小结】

- Model 1 和 Model 2 体系结构用于开发 Web 应用程序。
- 为了克服 Model 1 体系结构的安全性和维护问题，引入了 Model 2 体系结构。

- Model 2 体系结构也称为 MVC，它包含模型、视图和控制器 3 个组件。
  - 模型负责实际的数据处理，如数据库连接，请求数据库和实现业务逻辑。
  - 视图是客户端应用程序的图形数据表示，与实际数据处理无关。
  - 控制器对鼠标或键盘输入做出响应，以命令模型和视图对象进行相应的更改。

【单元自测】

1. (　　)页面在 Model 1 体系结构中负责处理请求。

   A. XML　　B. DHTML　　C. JSP　　D. HTML

2. Model 2 体系结构也称为(　　)体系结构。

   A. 模型—视图—控制器　　B. 视图—模型—控制器

   C. 视图—控制器—模型　　D. 控制器—模型—视图

3. Model 2 体系结构的模型对象被编写为(　　)。

   A. Applet　　B. JSP 页面　　C. JavaBean　　D. JSP 表达式

4. Model 2 体系结构的(　　)对象向客户端呈现应用程序界面。

   A. 模型　　B. 控制器　　C. 视图　　D. Servlet

5. Model 2 体系结构的控制器对象是一个(　　)。

   A. JSP 页面　　B. HTML 页面　　C. Servlet　　D. JavaBean

【上机实战】

上机目标

- 掌握模型组件、视图组件和控制器组件。
- 掌握 MVC 模式在 Web 程序中的应用。

上机练习

◆ 第一阶段 ◆

练习 1：使用理论课中的简单框架实现网上书店的功能

【问题描述】

在前面章节中，使用 Servlet 实现了对数据库中 books 表进行添加书籍、按某个编号删除书籍、修改某个书籍的信息和查询书籍相关信息的操作。本例中，使用理论课中的框架来实现同样的功能。

【问题分析】

(1) 使用三层结构完成数据库操作，使用业务类完成数据库 books 表的字段的更新操作。

(2) 添加书籍、查询所有书籍信息、按某个编号删除书籍、修改某个书籍的信息是 4 个不同的业务，分别由 AddBookAction、ShowAllBooksAction、DelBookAction 和 UpdateBookAction这 4 个业务的Action来完成，并且这 4 个业务类都实现了Action接口，并重写了接口中定义的方法。

(3) 需要建立 5 个页面，分别是：index.jsp 页面实现功能选择；addBookPage.jsp 页面实现用户输入书籍的详细信息；showAllBooksPage.jsp 页面实现显示数据库里所有书籍信息；delBookPage.jsp 页面实现用户输入要删除的书籍编号；updateBookPage.jsp 实现用户输入要修改的书籍信息。

(4) 需要建立Properties配置文件来定义这4个Action请求的URL和处理完业务后转到的页面。

【参考步骤】

(1) 使用单元六中上机部分第二阶段练习 2 的部分代码，在数据库的基础类 BookDAO.java 中增加两个抽象方法，完成新书的增加和删除功能，代码如示例 11-16 所示。相应的实现类 BookDAOImpl.java 也进行如示例 11-17 所示的修改。

示例 11-16：

```
package dao;

import java.util.*;
import bean.BookBean;

public abstract class BookDAO    extends BaseDAO{
    public abstract List<BookBean> getAllBooks() ;
    public abstract int updateBook(BookBean bookBean);
    public abstract BookBean getBookById(Integer id);
    public abstract int insertBook(BookBean bookBean) ;
    public abstract int deleteBookById(Integer id);
}
```

示例 11-17：

```
package dao;

import java.sql.PreparedStatement;
import java.sql.SQLException;
import java.util.ArrayList;
import java.util.Iterator;
import java.util.LinkedHashMap;
import java.util.List;
```

```
import java.util.Map;

import support.DBCommand;
import bean.BookBean;

public class BookDAOImpl extends BookDAO {
    private PreparedStatement pstm;

    // 获取所有的图书
    public List<BookBean> getAllBooks() {
        List<BookBean> bookList = new ArrayList<BookBean>();
        try {
            pstm = this.getConn().prepareStatement("SELECT * FROM books");
            List<Map<String, Object>> list =
                DBCommand.execQuery(pstm);
            // 迭代
            Iterator<Map<String, Object>> it = list.iterator();
            while (it.hasNext()) {
                Map<String, Object> row = it.next();
                BookBean bookBean = new BookBean();
                bookBean.setBookId(
                    new Integer(row.get("bookid").toString()));
                bookBean.setIsbn(row.get("isbn").toString());
                bookBean.setName(row.get("name").toString());
                bookBean.setAuthor(row.get("author").toString());
                bookBean.setPrice(
                    new Float(row.get("price").toString()));
                bookBean.setStock(
                    new Integer(row.get("stock").toString()));
                bookList.add(bookBean);
            }
        } catch (Exception e) {
            e.printStackTrace();
        }
        return bookList;
    }

    // 修改图书信息
    public int updateBook(BookBean bookBean) {
        try {
        pstm = this.getConn().prepareStatement("update books set isbn
                = ?,name=?,author=?,price=?,stock = ? where bookid=?");
        Map<Object, Object> paramsMap = new LinkedHashMap<Object, Object>();
            paramsMap.put("isbn", bookBean.getIsbn());
            paramsMap.put("name", bookBean.getName());
            paramsMap.put("author", bookBean.getAuthor());
            paramsMap.put("price", bookBean.getPrice());
```

```
            paramsMap.put("stock", bookBean.getStock());
            paramsMap.put("bookid", bookBean.getBookId());
            return DBCommand.execUpdate(pstm, paramsMap);
        } catch (Exception e) {
            e.printStackTrace();
        }
        return -1;
    }

    // 根据 bookid 查找图书
    public BookBean getBookById(Integer id) {
        BookBean bookBean = null;
        try {
        pstm = this.getConn().prepareStatement(
                    "SELECT * FROM books WHERE bookid = ?");
        Map<Object, Object> paramsMap = new LinkedHashMap<Object, Object>();
            paramsMap.put("bookid", id.intValue());
            List<Map<String, Object>> stuList =
                    DBCommand.execQuery(pstm,paramsMap);
            if (stuList.size() != 0) {
                bookBean = new BookBean();
                Map<String, Object> row = stuList.get(0);
                bookBean.setBookId(
                    new Integer(row.get("bookid").toString()));
                bookBean.setIsbn(row.get("isbn").toString());
                bookBean.setName(row.get("name").toString());
                bookBean.setAuthor(row.get("author").toString());
                bookBean.setPrice(
                    new Float(row.get("price").toString()));
                bookBean.setStock(
                    new Integer(row.get("stock").toString()));
            }
        } catch (Exception e) {
            e.printStackTrace();
        }
        return bookBean;
    }

    //增加新书到数据库
    public int insertBook(BookBean bookBean) {
        int res = -1;
        try {
            pstm = this.getConn().prepareStatement(
                    "insert into books values(?,?,?,?,?)");
        Map<Object, Object> paramsMap = new LinkedHashMap<Object, Object>();
            paramsMap.put("isbn", bookBean.getIsbn());
            paramsMap.put("name", bookBean.getName());
```

```
            paramsMap.put("author", bookBean.getAuthor());
            paramsMap.put("price", bookBean.getPrice());
            paramsMap.put("stock", bookBean.getStock());
            res = DBCommand.execUpdate(pstm, paramsMap);
        } catch (SQLException e) {
            e.printStackTrace();
        }
        return res;
    }

  //根据 bookid 删除书籍
    public int deleteBookById(Integer id) {
        int res = -1;
        try {
            pstm = this.getConn().prepareStatement(
                    "delete from    books where bookid = ?");
        Map<Object, Object> paramsMap = new LinkedHashMap<Object, Object>();
            paramsMap.put("bookid", id);
            res = DBCommand.execUpdate(pstm, paramsMap);
        } catch (SQLException e) {
            e.printStackTrace();
        }
        return res;
    }
}
```

(2) 修改业务处理接口 BookService.java 和其实现类 BookServiceImpl.java，代码如示例 11-18、示例 11-19 所示。

示例 11-18：

```
package service;
import java.util.List;
import bean.BookBean;
public interface BookService{
    public abstract List<BookBean> getAllBooks() ;
    public abstract boolean updateBook(BookBean bookBean);
    public abstract BookBean getBookById(Integer id);
    public abstract boolean insertBook(BookBean bookBean) ;
    public abstract boolean deleteBookById(Integer id);
}
```

示例 11-19：

```
package service;
import java.util.List;
import bean.BookBean;
```

```
import dao.BookDAO;
import dao.BookDAOImpl;
import support.DBHelper;
public class BookServiceImpl implements BookService {

    private BookDAO dao = new BookDAOImpl();

    public BookBean getBookById(Integer id) {
        BookBean book = null;
        try {
            dao.setConn(DBHelper.getConn());
            book= dao.getBookById(id);
        } catch (RuntimeException e) {
            e.printStackTrace();
        }finally{
            DBHelper.closeConn();
        }
        return book;
    }

    public List<BookBean> getAllBooks() {
        List<BookBean> bookList = null;
        try {
            dao.setConn(DBHelper.getConn());
            bookList = dao.getAllBooks();
        } catch (RuntimeException e) {
            e.printStackTrace();
        }finally{
            DBHelper.closeConn();
        }
        return bookList;
    }

    public boolean updateBook(BookBean bookBean) {
         boolean res = false    ;
         try {
            dao.setConn(DBHelper.getConn());
              res = dao.updateBook(bookBean) >0 ? true :false ;
        } catch (RuntimeException e) {
            e.printStackTrace();
        }finally{
            DBHelper.closeConn();
        }
        return res ;
    }

    public boolean deleteBookById(Integer id) {
```

```
        boolean res = false   ;
         try {
            dao.setConn(DBHelper.getConn());
             res = dao.deleteBookById(id) >0 ? true :false ;
        } catch (RuntimeException e) {
            e.printStackTrace();
        }finally{
            DBHelper.closeConn();
        }
        return res ;
    }

    public boolean insertBook(BookBean bookBean) {
        boolean res = false   ;
         try {
            dao.setConn(DBHelper.getConn());
             res = dao.insertBook(bookBean)>0 ? true :false ;
        } catch (RuntimeException e) {
            e.printStackTrace();
        }finally{
            DBHelper.closeConn();
        }
        return res ;
    }
}
```

(3) 4 个 Action 分别如示例 11-20、示例 11-21、示例 11-22 和示例 11-23 所示。

示例 11-20：

```
package biz;

import javax.servlet.ServletContext;
import javax.servlet.http.HttpServletRequest;
import model.Action;
import model.ActionForward;
import bean.BookBean;
import service.BookService;
import service.BookServiceImpl;

public class AddBookAction implements Action {

    public ActionForward execute(HttpServletRequest request,
            ServletContext context) {
        String isbn = request.getParameter("isbn");
        String name = request.getParameter("name");
        String author = request.getParameter("author");
        String pr = request.getParameter("price");
```

```
            String qu = request.getParameter("stock");
            float   price = Float.parseFloat(pr);
            int stock = Integer.parseInt(qu);
            BookBean book = new BookBean();
            book.setIsbn(isbn);
            book.setName(name);
            book.setAuthor(author);
            book.setPrice(price);
            book.setStock(stock);
            //生成业务处理实例对象
            BookService service = new BookServiceImpl();
            boolean success = service.insertBook(book);
            ActionForward af = new ActionForward();
            if(success){
                af.setName("add");
            }
          return af;
    }
}
```

示例 11-21：

```
package biz;

import javax.servlet.ServletContext;
import javax.servlet.http.HttpServletRequest;
import service.BookService;
import service.BookServiceImpl;
import model.Action;
import model.ActionForward;
import bean.BookBean;
import java.util.List;

public class ShowAllBooksAction implements Action {

    public ActionForward execute(HttpServletRequest request,
            ServletContext context) {
        // 生成业务处理实例对象
        BookService service = new BookServiceImpl();
        List<BookBean> books = service.getAllBooks();
        request.setAttribute("list",books);
        ActionForward af = new ActionForward();
        af.setName("show");
        return af;
    }
}
```

示例 11-22：

```
package biz;

import javax.servlet.ServletContext;
import javax.servlet.http.HttpServletRequest;
import model.Action;
import model.ActionForward;
import service.BookService;
import service.BookServiceImpl;

public class DelBookAction implements Action {

    public ActionForward execute(HttpServletRequest request,
    ServletContext context) {
        String bookid = request.getParameter("bookid");
        // 生成业务处理实例对象
        BookService service = new BookServiceImpl();
        boolean flag = service.deleteBookById(new Integer(bookid));
        ActionForward af = new ActionForward();
        if (flag) {
            af.setName("delete");
        }
        return af;
    }
}
```

示例 11-23：

```
package biz;

import javax.servlet.ServletContext;
import javax.servlet.http.HttpServletRequest;
import service.BookService;
import service.BookServiceImpl;
import model.Action;
import model.ActionForward;
import bean.BookBean;

public class UpdateBookAction implements Action {

    public ActionForward execute(HttpServletRequest request,
    ServletContext context) {
        String   bookid = request.getParameter("bookid");
        String   price = request.getParameter("price");
        String   stock = request.getParameter("stock");
        // 生成业务处理实例对象
```

```
            BookService service = new BookServiceImpl();
            //先根据 isbn 找到书
            BookBean book =     service.getBookById(new Integer(bookid));
            //再修改
              book.setPrice(new Float(price));
              book.setStock(new Integer(stock));
              //更新书的数量和价格
            boolean flag = service.updateBook(book);
            ActionForward af = new ActionForward();
            if(flag){
                af.setName("update");
            }
            return af;
        }
    }
```

(4) 5 个 JSP 页面的代码如示例 11-24、示例 11-25、示例 11-26、示例 11-27 和示例 11-28 所示。

示例 11-24：

```
<%@ page contentType="text/html; charset=utf-8"%>
<html>
  <head> <title>请选择功能菜单</title></head>
<body bgcolor="#ffffff">
<center>
  <h4>请选择书籍的功能菜单</h4>
  <table>
    <tr>
      <td><a href="addBookPage.jsp">添加</a></td>
      <td><a href="doShowAll.action">查询</a></td>
      <td><a href="editBookPage.jsp">修改</a></td>
      <td><a href="delBookPage.jsp">删除</a></td>
    </tr>
  </table>
</center>
</body>
</html>
```

示例 11-25：

```
<%@ page contentType="text/html; charset=utf-8" %>
<%@taglib uri="http://java.sun.com/jsp/jstl/core" prefix="c"%>
<html>
<head><title>显示所有书籍页面</title></head>
<jsp:include page="index.jsp" flush="true"/>
<body bgcolor="#ffffff">
```

```
<center>
  <h4>所有书籍信息</h4>
  <table border="1" cellpadding="3" cellspacing="3">
        <tr>
          <th>编号</th><th>ISBN</th><th>书名</th><th>作者</th>
          <th>价钱</th><th>库存</th>
        </tr>
        <c:forEach var="bookbean" items="${requestScope.list}">
          <tr>
         <td>${bookbean.bookId}</td><td>${bookbean.isbn}</td>
<td>${bookbean.name}</td><td>${bookbean.author}</td>
<td>${bookbean.price}</td><td>${bookbean.stock}</td>
          </tr>
        </c:forEach>
      </table>
</center>
</body>
</html>
```

示例 11-26：

```
<%@page contentType="text/html; charset=utf-8"%>
<html>
    <head><title>添加书籍页面</title></head>
    <jsp:include page="index.jsp" flush="true"/>
    <body bgcolor="#ffffff">
        <center>
            <h4>请输入数据的详细信息</h4>
            <form name="form1" action="doAdd.action" method="POST">
                <table>
                <tr><td>编号 <input type="text" name="isbn" /></td></tr>
                <tr><td>书名 <input type="text" name="name" /></td></tr>
                <tr><td>作者 <input type="text" name="author" /></td></tr>
                <tr><td>价钱 <input type="text" name="price" /></td></tr>
                <tr><td>数量 <input type="text" name="stock" /></td></tr>
                <tr><td><input type="submit" value="添加" />
                    <input type="reset" value="清空" /></td></tr>
                </table>
            </form>
        </center>
    </body>
</html>
```

示例 11-27：

```
<%@page contentType="text/html; charset=utf-8"%>
<html>
```

```
<head><title>删除书籍页面</title></head>
<jsp:include page="index.jsp" flush="true"/>
<body bgcolor="#ffffff">
<center>
  <h4>删除书籍</h4>
  <form name="form1" action="doDelete.action" method="POST">
    <table>
      <tr><td>编号 <input type="text" name="bookid"/></td></tr>
      <tr><td><input type="submit" value="删除"/>
          <input type="reset" value="清空"/></td></tr>
    </table>
  </form>
</center>
</body>
</html>
```

示例 11-28：

```
<%@page contentType="text/html; charset=utf-8"%>
<html>
   <head><title>修改书籍信息页面</title></head>
   <jsp:include page="index.jsp" flush="true"/>
   <body bgcolor="#ffffff">
   <center>
   <h4>修改书籍的价格和库存</h4>
   <form name="form1" action="doUpdate.action" method="POST">
   <table>
   <tr><td>编号 <input type="text" name="bookid" /></td></tr>
   <tr><td>价钱  <input type="text" name="price" /></td></tr>
   <tr><td>数量  <input type="text" name="stock" /></td></tr>
   <tr><td><input type="submit" value="修改" />
         <input type="reset" value="清空" /></td></tr>
   </table>
   </form>
   </center>
   </body>
</html>
```

(5) 建立 Properties 配置文件，代码如示例 11-29 所示。

示例 11-29：

```
doAdd=/doAdd,biz.AddBookAction,add,doShowAll.action
doDelete=/doDelete,biz.DelBookAction,delete,doShowAll.action
doShowAll=/doShowAll,biz.ShowAllBooksAction,show,showAllBooksPage.jsp
doUpdate=/doUpdate,biz.UpdateBookAction,update,doShowAll.action
```

(6) 部署该应用到服务器，验证各个功能。

注意，鉴于篇幅，这个网上书店的应用有以下不完善的地方。

- 页面里表单数据没有做数据验证。
- 添加书籍业务默认是数据库中不存在要添加的数据，实际上添加书籍的业务由多个基本业务组成。先判断要添加的书籍在数据库中是否存在，如果存在，只是做数量上的增加；如果价格和原价格不同，使用哪个价格？同样删除业务和更新业务都是默认数据库已经存在书籍的情况，实际的业务也不是这样的。
- 理论课中讲过这个框架的缺陷，在实际应用中，每个 Action 默认都是成功的，然后控制权转向到成功后的页面。实际上，业务的结果有多种，例如添加书籍，如果成功，控制权转到 showAllBooksPage.jsp 页面；如果插入不成功，提示用户插入书籍有误，则转到预定义的 error.jsp 页面。针对不同情况，页面转到不同的结果页面。这个配置文件如何写？PropertyUtil 工具类如何解析配置文件呢？练习 2 就凸显了这个问题，用户登录的结果有两种，成功或失败。

◆ 第二阶段 ◆

练习 2：给网上书店添加登录功能

【问题描述】

改进练习 1 中提到的不完善的地方，并给网上书店系统增加登录功能，查询数据库中的 UserInfo 表。用户分为普通用户和超级用户，普通用户只能够查询书籍的详细信息，超级用户可以修改和删除书籍。

【问题分析】

(1) 与数据库 UserInfo 表交互，需要建立与数据库交互的操作类 UserInfoDAO，类中定义了查询用户的方法，方法的入参是 UserInfo 实体类对象。

(2) 设计检查用户是否是超级用户的业务类 CheckUserInfoAction，这个业务类实现了 Action 接口，并重写了接口中定义的方法。

(3) 需要重新设计 Properties 配置文件来实现用户请求的 URL 和业务类的映射，以及处理完业务后根据不同的结果转到不同的页面。

(4) 需要重新设计 PropertyUtil 类来解析配置文件。

【拓展作业】

使用理论课中的 MVC 框架实现银行 ATM 柜员机系统，要求能够实现存钱、取钱、查询余额、转账、修改用户密码和查询交易记录功能。(数据库需要 Account 表和 Record 表，其中 Account 表用来存放用户的账户信息，Record 表用来存放交易记录。存钱或取钱都向 Record 表插入一条交易记录，转账则向 Record 表插入两条记录，修改密码和查询则不向 Record 表插入记录。)

参考文献

[1] 王春明、史胜辉. JSP Web 技术及应用教程[M]. 2 版. 北京：清华大学出版社，2018.

[2] 马建红，等. JSP 应用与开发技术[M]. 3 版. 北京：清华大学出版社，2019.

[3] 谷志峰. JSP 程序设计实例教程[M]. 北京：电子工业出版社，2017.